Decision Engineering

AF779957

Series Editor

Professor Rajkumar Roy
Department of Enterprise Integration
School of Industrial and Manufacturing Science
Cranfield University
Cranfield
Bedford
MK43 0AL
UK

Other titles published in this series

Cost Engineering in Practice
John McIlwraith

IPA – Concepts and Applications in Engineering
Jerzy Pokojski

Strategic Decision Making
Navneet Bhushan and Kanwal Rai

Product Lifecycle Management
John Stark

From Product Description to Cost: A Practical Approach
Volume 1: The Parametric Approach
Pierre Foussier

From Product Description to Cost: A Practical Approach
Volume 2: Building a Specific Model
Pierre Foussier

Decision-Making in Engineering Design
Yotaro Hatamura

Composite Systems Decisions
Mark Sh. Levin

Intelligent Decision-making Support Systems
Jatinder N.D. Gupta, Guisseppi A. Forgionne and Manuel Mora T.

Knowledge Acquisition in Practice
N.R. Milton

Global Product: Strategy, Product Lifecycle Management
and the Billion Customer Question
John Stark

Enabling a Simulation Capability in the Organisationon
Andrew Greasley

Network Models and Optimization
Mitsuo Gen, Runwei Cheng and Lin Lin

Gudela Grote

Management of Uncertainty

Theory and Application in the Design of Systems and Organizations

Springer

Professor Gudela Grote
ETH Zürich
Department of Management, Technology, and Economics
Kreuzplatz 5
8032 Zürich
Switzerland
ggrote@ethz.ch

ISSN 1619-5736
ISBN 978-1-4471-2510-5 e-ISBN 978-1-84882-373-0
DOI 10.1007/978-1-84882-373-0
Springer Dordrecht Heidelberg London New York

British Library Cataloguing in Publication Data
A catalogue record for this book is available from the British Library

© Springer-Verlag London Limited 2009
Softcover reprint of the hardcover 1st edition 2009
Apart from any fair dealing for the purposes of research or private study, or criticism or review, as permitted under the Copyright, Designs and Patents Act 1988, this publication may only be reproduced, stored or transmitted, in any form or by any means, with the prior permission in writing of the publishers, or in the case of reprographic reproduction in accordance with the terms of licences issued by the Copyright Licensing Agency. Enquiries concerning reproduction outside those terms should be sent to the publishers.
The use of registered names, trademarks, etc. in this publication does not imply, even in the absence of a specific statement, that such names are exempt from the relevant laws and regulations and therefore free for general use.
The publisher makes no representation, express or implied, with regard to the accuracy of the information contained in this book and cannot accept any legal responsibility or liability for any errors or omissions that may be made.

Cover design: eStudioCalamar, Figueres/Berlin

Printed on acid-free paper

Springer is part of Springer Science+Business Media (www.springer.com)

Preface

As I write, the financial systems of the world are collapsing with still no clear indication of what the consequences will be and which measures should be taken to avoid such a crisis in the future. There seems to be agreement though, that the financial instruments introduced in the past few decades entailed far too much complexity and uncertainty and that there was too little regulatory control over the use of these instruments.

Management of uncertainty with the aim of achieving self-control is the core concern of this book. It was not written with a focus on financial systems, but many concepts developed in this book are applicable to this field as well. The generic principles of reducing, maintaining or increasing uncertainties in view of the different contingencies an organization is faced with, the fundamental issue of how much control is possible and who should be in control, and the question of how much and what kind of regulation is necessary with the overall aim of finding an appropriate balance between system stability and flexibility are at the centre of heated debates on the future of finance.

The financial crisis is also a very drastic example of the multi-layered nature of the problems discussed in this book. The management of uncertainty concerns all of us as individuals, be it as operational or strategic decision-makers in organizations, as private persons managing our lives, or as citizens involved in the governance of risk on a societal level. The management of uncertainty is also one of the main tasks of public and private organizations on a strategic level as well as in day-to-day operations. In the case of the financial system, it seems that for quite some time problems were located mainly on the individual level. The losses produced by Nick Leeson for Barings Bank or by Jérôme Kerviel for the Société Générale were attributed to personal greed and incompetence. The organizational and societal dimension was only gradually discovered as it became clear that even the most competent and conscientious individuals were not able to handle the uncertainties involved in investment banking adequately. Rich people becoming ever richer through sophisticated exploitation of the uncertainties deliberately built into new financial instruments were celebrated as a victory of capitalism. The forces of the free financial markets were to be left unabated by state regulation in order to maximize this exploitation. As state intervention sets in, we all get to pay for the riches of those chosen few. More regulation and maybe even a tamed form of capitalism seem to be regarded as necessary next steps in the attempt to find a new and better balance between financial stability and flexibility. In a small way, this book tries to help people who have to make decisions on managing uncertainties in order to balance stability and flexibility, be it in finance or any other realm of the world of work.

In turn, many people have helped me tremendously to write this book. The initiative came from Springer when Anthony Doyle asked me whether I might be interested in expanding the ideas I had expressed on the management of uncertainty in system design in a short article, which appeared in 2004 in Annual Reviews in Control. However much I have regretted my positive response to him in those times when the book did not seem to materialize at all, I am very grateful to him for having posed this challenge to me and for having been incredibly patient with its slow progress. My sincere thanks also go to Claire Protherough at Springer UK for her unwavering help in the final stages of editing the book.

The ideas presented in this book had started to take shape long before I ever sat down to write it. Many people have contributed to this process, foremost my collaborators over the years: Cuno Künzler, Steffen Weik, Toni Wäfler, Martina Zölch, Cornelia Ryser, Anna Windischer, Barbara Klampfer, Sven Grund, Sabine Raeder, Sylvia Manchen Spörri, Scott Little, Lille Springall, Alexandra Totter, Enikö Zala-Mezö, Hannes Günter, Anette Wittekind, Barbara Künzle, Marius Gerber, Daniel Boos, Michaela Kolbe, Johann Weichbrodt, and Cécile Tschopp. Chapters 4 to 8 build extensively on work carried out and published together with some of them. Frequent discussions on how to conceptually unite the various endeavours in our research group were an important motivation and inspiration for developing the management of uncertainty framework presented in the first three and final chapters.

I also owe many thanks to colleagues in various parts of the world for helping me to form and sharpen my arguments, including Michael Arthur, Christof Baitsch, Guy Boy, Rob Briner, Chris Clegg, Rhona Flin, David Guest, Tim Hall, Erik Hollnagel, Larry James, Bruno Staffelbach, Johan Stahre, Franziska Tschan, Nick Turner, Eberhard Ulich, John Wilson, Sid Winter and Yan Xiao. And there were five absolutely invaluable colleagues who read through the whole book and praised and criticized with the perfect balance needed for the self-assured self-criticism that spurred on the final revision: Toby Wall, Georg von Krogh, Andrew Hale, Carlo Cacciabue, and Hannes Günter – thank you! For transforming the manuscript into a proper book, my thanks go to Jacqueline Hohermuth and Mareike Haase.

This book could not have been written without the many contacts and collaborations I have had over the years with practitioners from a variety of public and private organizations. Many thanks to all of them for sharing their experiences and for being open to "academic" dialogue. In various parts of the book, examples from the railways are given – which is by no means accidental because I love to travel by train and because I had the opportunity to work with some exceptional people at the Swiss Railways, who were eager to try out new approaches to managing uncertainty. My thanks go to Andreas Hönger and Hans Vogt especially. I also wish to thank Ueli Straub, Stephan Müller and Ernst Zirngast at Swiss Reinsurance Company for the long and very fruitful cooperation on the assessment of safety management and safety culture, which provided much of the material presented in Chapter 7.

Finally, a thousand thanks and more to the three most important men in my life: Liam and Alan for stirring up my life with many new uncertainties, and Tom for being my first and last critic and for being the source of both crucial uncertainties and certainties in my life.

Zurich, January 2009

Gudela Grote

Contents

1

Introduction

Most of the time, we like things to happen in a predictable way. We plan our business ventures, our families, and our holidays and we want reality to follow the plan. Uncertainties associated with the availability of resources, the occurrence of events along the way, and ultimately with goal achievement are undesired inconveniences, tolerated at best. But then there are the odd occasions when we want to be surprised, to be held in suspense, to discover the unknown. It seems, especially in Western societies, that there is a growing tendency to oscillate between these two attitudes towards uncertainty. While for the most part, we wish to control our lives and our environment in ever increasing detail and to cover the uncontrollable by insurances of various kinds, at certain times we seek adventure in ever more extreme forms. The financial crisis – unfolding as the writing of this book was in its final stages – seems to indicate that a few people have made a very good living out of their need for adventure, while the rest of us suffer, as our economic and psychological foundations of certainty and control are shaken.

In organizations, there is even more reason to avoid uncertainties as they may endanger business success. But at the same time, uncertainties are sought out as sources of innovation, and the ability to flexibly handle uncertainties becomes a competitive advantage. The importance of uncertainty for organizational functioning has long been recognized: "Uncertainty appears as the fundamental problem for complex organizations, and coping with uncertainty as the essence of the administrative process" (Thompson 1967, p. 159). In organization theory and strategic management, frameworks for contingent use of different strategies for managing uncertainty have been developed. "Uncertainty about the firm's future is the core of strategy; it is the essential reason to engage in strategic planning (...) A first step toward making planning truly strategic will occur if top management calls for, and participates in, a no-nonsense consideration of uncertainty" (Allaire and Firsirotu 1989, p.16).

More recently, the main concern in management thinking has been to overcome dichotomies and to search for a middle ground between the extremes of either avoiding or embracing uncertainty as the sole strategy. Thus, the question becomes which uncertainties to avoid and which to embrace. Over the last century, with increasingly fast and global competition, an ever more clear understanding has developed that organizations are continuously faced with uncertainties from within and from without, which can only partially be reduced. How to live with uncertainties or even take advantage of them without losing the stability created by reducing uncertainty has become an important issue not only in research on organizations but also in practical management guidelines (*e.g.,* Brown and Eisenhardt 1997; O'Reilly and Tushman 2004). The fundamental objective is to balance stability and flexibility in accordance with internal and external demands, which requires addressing and managing strategic contradictions such as short-term performance and long-term adaptability or differentiation and integration (Smith and Tushman 2005). Recent research into extremely innovative companies has provided very interesting examples of how this balance can be achieved (*e.g.,* Benner and Tushman 2003; Brown and Eisenhardt 1997). The way Google is organized may serve as an illustration of the concurrent efforts to stabilize processes through introducing more structures and to keep processes as open as possible in order to stimulate innovation (see Example 1.1).

Example 1.1. Managing uncertainty at Google

Much has been written about the Google management style, which relies on a mixture of chaos and structure. While all functions apart from engineering are managed in a traditional, structured way, engineering is said to be governed by partially organized chaos. Teams which are deliberately kept small at 3 to 5 members, obtain maximum freedom in developing products with sometimes marginal specifications. There are a number of core instruments to be used during software development (project management, compiling, debugging, etc.) that are taught to all newcomers. These instruments provide a formalized, "bureaucratic" framework for the otherwise very open, innovation-focused process. All code is open to all members of Google, except for the famed ranking algorithm and the advertising key words. Therefore, anybody can initiate new ideas from (almost) any part of existing code. Everybody is expected to spend 20% of their time on their own ideas which may or may not develop into larger product development schemes. These schemes will then be staffed by development teams which would use the other 80% of their working time to implement them.

Google's creative spirit is supported by a general culture of coolness and innovation. People are hired for their brains and their "Googley" attitude and once they have passed the very high Google hiring bar, they are left to do their own thing. This high regard for employees and trust in their motivation and ability is not only apparent in the high decision latitude given to them, but is also demonstrated in the provision of a very comfortable and playful environment with the proverbial "free lunch", pool tables, massage chairs, and recreation zones.

Projects are run based on road maps which contain the rough direction to be taken, more or less over the following year. More detailed development plans are drawn up for three month periods, including the particular features to be delivered. Frequently, changes are required even within these quarter plans. Day-to-day coordination happens through short daily stand-up meetings and at least one weekly video-conference with other project team members across the world. Even software engineers with no formal management function spend up to 30 or 40% of their time coordinating project activities.

There are interlinked bottom-up and top-down processes for deciding which new products should be developed. It is mainly through the 20% projects that new ideas are brought in bottom-up. In order to structure innovation top-down, strategic decisions are made on whole product domains (*e.g.,* buying Youtube), or, more specific roadmaps for particular products and groups of products are developed. However master plans are still considered "uncool". Also, longstanding and more experienced members of the organization, including the founders themselves, try to place product ideas with engineers informally. Ideas get taken up by convincing others that these ideas have impact and are challenging, interesting and useful. One important task of engineering managers is to align individual initiatives with overall strategy.

From an uncertainty perspective, it is quite obvious that Google is faced with high amounts of external uncertainty, be it from new developments and use patterns in the internet or from the activities of direct competitors. It is also faced with considerable internal uncertainty regarding the definition of new products and the openness of the process that leads to these products. While much of the organization is geared towards coping with uncertainty or even increasing uncertainty, *e.g.,* by starting new product lines, there are some elements also aimed at reducing uncertainty, such as clear guidelines on how to write, test and document code. There is also an increasing awareness of the fact that innovation does not necessarily originate from maximum openness, but is partially a product of structured strategic decision-making. As a very interesting innovation in itself, former Stanford researcher and McKinsey consultant Shona Brown was hired with the explicit assignment of keeping structure and chaos in optimal balance at Google.

The Google example brings up some issues that will be at the centre of the following two chapters: how to identify the different kinds of uncertainty an organization is faced with and how to decide on the best way to handle these uncertainties. Specifically, which uncertainties should be reduced, maintained, or possibly even increased, in order to secure business success. When dealing with these questions, one may try to develop objective accounts of uncertainties and derive design strategies from considering various contingencies related to these uncertainties. At the same time, it is important to keep in mind that decision-makers in organizations enact their environment and the contingencies it contains through selective attention and interpretation. This is also true for the scientific endeavours that attempt to guide decision-makers: "organization theory not only reflects organizational reality, it also produces that reality. [...] The interplay of organization theories is in reality a contest over the future shaping of the organizational world" (Astley and Van de Ven 1983, p. 269-270). The more uncertainty there is on what is right or wrong in a certain situation, the more individual and collective enactment and sensemaking of the situation will take place in order to cope with insufficient and ambiguous information (Weick 1995). In this book, a rationalistic account of uncertainty management is coupled with the awareness of the impact sensemaking processes have on the strategic and operational choices made.

This double perspective is also an important element of decision-making itself and of leadership more generally. Uncertainty is generally unsettling to people because it frustrates the need for cognitive closure, meaning the desire for definite knowledge on whatever issue is at stake (Kruglanksi and Webster 1996). Managers, like any other individual, may therefore try to reduce uncertainty prematurely, not take advantage of exploration due to the ensuing increase in uncertainty, or not be willing to face uncertainty at all. These tendencies will strongly influence the perception of uncertainty and the presumably rational choice for a particular course of action.

Moreover, demands for tolerating and actively managing uncertainty are increasing due to the requirement to match the complexity of organizational reality with equally complex portfolios of leadership behaviour, for instance, by switching between directive and participative leadership styles (Denison *et al.* 1995; Hunt *et al.* 2004). A similar idea is expressed by the notion of mastery of opposites, which concerns adaptive use of directive and enabling as well as strategic and operational leadership. The ability for mastering these opposites in ways adapted to situational and employee needs has been found to be strongly linked to team effectiveness (Kaiser *et al.* 2007). Also, earlier work by Bourgeois and Eisenhardt (1988) on IT companies has shown that successful companies are generally better able to manage paradox such as fast and careful decision-making, a powerful CEO in a powerful top management team, or risky innovation combined with safe implementation.

The double perspective taken in this book, that is a rationalistic account of uncertainty management combined with acknowledging the impact of sensemaking processes, lends itself to what has been called a configurational approach to organizational analysis (Meyer *et al.* 1993). Building on the insights from contingency theory, which has helped to clarify links between environment, structure, work processes, and technology, a configurational approach attempts to create rich descriptions of reciprocal and multivariate relationships among individual, group, organizational and environmental characteristics. Such an approach renders the identification of clear-cut, cause-and-effect relationships, which could be used to control situations for maximum success, much more difficult. However, it furthers an understanding of organizational reality, where complexity matches the complexity of the situation encountered, which is a fundamental requirement for effective action (Ashby 1956; Weick 1976).

A final point to make about the basic conception of the book concerns the chosen focus on uncertainty. Many people with whom I have discussed the plan for this book felt that risk would be a more appropriate focus, particularly in view of increasing worries about financial, technological and environmental risks getting out of control. While these worries are explicitly addressed in several chapters of the book, the overall framework centres on uncertainty and not risk. Uncertainty is considered to be the more generic concept, which can easily be linked to risk in its most basic form as an uncertain event or in more specific definitions such as the product of probability and damage. Uncertainty in terms of insufficient or ambiguous knowledge about cause-and-effect relationships can be regarded as the "neutral" source of risk because the predictability and transparency of the situation are reduced and thereby also the control needed to avoid damage and realize opportunities. When uncertainties are managed well, a substantial prerequisite for good risk management is established.

1.1 Intended Readership

The book has been written with both practitioners and researchers in mind, even though overall, it is geared more towards academics. From a practitioner's perspective, the most useful elements of the book should be the tools described in the different chapters, mainly the overall framework for analyzing uncertainty and deciding on ways to manage uncertainty, and more specifically, the typology for organizational rules, the criteria and design heuristic for decisions on automation, or instruments for assessing collaborative planning, safety culture, and career orientations. Given the broad range of issues discussed in relation to system and organizational design, practitioners from a variety of professional backgrounds will find relevant material in the book. These may be strategic decision-makers, as well as line managers and professionals in technical fields, or human resource managers. Beyond the specific tools introduced the book aims to further the broader understanding of issues related to the management of uncertainty in a variety of settings and from different actors' view points. This may facilitate reflection on the effects of one's own decisions taken in the past and prepare for taking decisions in the future in fuller recognition of their impact on the management of uncertainty.

For researchers, the book aims to integrate very diverse research streams into an overall framework for understanding uncertainty, its relevance in organizations, and different approaches to its management. Furthermore, the comprehensive review of current research and thinking in several domains of system and organizational design, along with suggested new conceptual developments in these domains, should be a useful guide for future research. Some specific suggestions for possible roads to take in research are presented in the final chapter.

Finally, students in the various domains discussed in the book should find the introduction to core questions in the different fields illuminating. For them, the book might primarily serve as a map to guide their search for personal research questions and as a source of relevant literature.

1.2 Outline of the Book

The book is organized around an analytical framework for understanding uncertainty in organizations. This framework is developed in Chapter 2 and includes causes, sources, contents, and consequences of uncertainty, as well as different actors involved in the management of uncertainty. It centres on the uncertainties' effects for controlling action towards achieving certain outcomes. Securing control is taken to be the ultimate goal of managing uncertainty. The framework's elements can be applied to analyze conditions of managing uncertainty from a strategic and an operational perspective and examples for both are provided. Moreover, the framework allows for the adoption of a rationalistic, objective stance on managing uncertainty, and also the consideration of the impact of individual and collective enactment and sensemaking, which is of particular importance precisely in situations with high levels of uncertainty. While there is some reference to personal factors involved in managing uncertainty, such as risk propensity or personal values and beliefs, the main focus is on organizational processes and their determinants and consequences. The overall aim of the analytic framework is to help decision-makers in organizations to sketch an uncertainty landscape that supports informed decisions on managing these uncertainties so as to achieve desired outcomes.

In Chapter 3, generic principles for managing uncertainty are derived from existing research and a decision process is suggested which aids the choice between reducing, maintaining or increasing uncertainty in order to reach particular goals. No particular "theory of the firm" was chosen as a guiding principle, but rather elements of different theoretical approaches were integrated in as much as they concern the management of uncertainty. For instance, the principal-agent theory points out problems related to the transfer of uncertainty between principal and agent, while the resource based view stresses uncertainties stemming from the dependence on specific external and internal resources. The chapter begins by contrasting minimizing versus coping with uncertainties as generic approaches whose suitability depends on various contingencies, most of all the amount and kinds of uncertainty with which an organization is faced. In accordance with current thinking in strategic management and organization theory, the balance between stability and flexibility forms the centre of all further considerations of managing uncertainties well. The concepts of loose coupling, enabling bureaucracy, semistructures and ambidextrous organizations are discussed with respect to their theoretical and practical value for balancing stability and flexibility. Subsequently, a decision process is proposed that supports systematic consideration of the costs and benefits of reducing, maintaining or increasing uncertainty in a particular context, as well as of the individual and collective belief systems that influence the perception and assessment of these costs and benefits.

This decision process is taken up again as appropriate in the following chapters when different domains of organizational functioning are discussed in more detail. The choice of certain domains was motivated by two considerations; firstly, the domains are not only of high relevance to company success but also touch on broader societal concerns; secondly, the domains can serve as good illustrations of the analytic framework developed at the outset of the book. Specifically, coordination, planning, automation, safety, and human resource management were chosen as domains. While the first four are mainly discussed from an organizational perspective, the latter permits perspectives to be switched and contrasts an individual view on the management of uncertainty with that of an organization. Coordination and planning can be considered as core to managing uncertainty and directly affect sources and contents of uncertainty, such as insufficient or ambiguous information regarding particular undesired states and the courses of action to be taken to change to more desired states. Automation is discussed from the viewpoint of human control over technical systems, thereby focusing on the effects of uncertainty regarding the overall objective of managing uncertainty, which is securing control. Subsequently, safety is introduced as an important consequence of how well uncertainty is managed. The discussion of safety management illustrates well how assumed consequences of a chosen course of action impact decisions on reducing, maintaining or increasing uncertainty. Finally, different actors and their different perspectives on uncertainty are introduced when discussing the employment relationship. Naturally, the views of different groups of actors are also touched upon in other chapters, but the focus on the employment relationship allows a more systematic contrast between personal and organizational perspectives. Some more detail on the chapters follows.

In Chapter 4 coordination in organizations is discussed with the specific focus on rules and routines as coordination mechanisms. From a contingency perspective, the issue is to establish a balance between stability and flexibility using rules and routines that allow coping with uncertainty while at the same time providing sufficient standardization to ease coordination demands. Based on newer research on organizational routines, it is argued that routines can be sufficiently flexible to support adaptation in unforeseen situations, especially when they are backed by flexible rules. This argument leads to a more sophisticated perspective on rules and standardization. Instead of only taking the amount of rules as an indication of more or less standardization, types of rules need to be studied. One way of distinguishing rule types is based upon the level of action regulation targeted; that is goal, process or concrete action. An example of the analysis of a railway company's rulebook is presented which shows in particular a mismatch between rule types, uncertainties to be handled and actors' competence profiles. Such analyses can serve as input into the redesign of rules aimed at creating flexible routines capable of absorbing uncertainty when needed. Systematic rule management can, at least partially, resolve the dilemma stemming from the contradictory requirements on coordination through rules: reducing demands on cognitive and organizational resources while at the same time freeing resources for response flexibility and adaptability.

In Chapter 5, the focus is on planning as a fundamental component in managing uncertainty, given that the very nature of planning is to determine action paths into an uncertain future. Specifically, the discussion centres on operational planning within and across organizations, which relates to internal logistical processes and external supply relationships. A number of problems in operational planning are identified, especially tensions between central and decentralized planning, autonomy and interdependence, and more and less powerful actors. Some caution is expressed as regards the general utility of planning and rationalistic beliefs in planning as a blueprint for action. It is argued that planning should be regarded as a resource for situated action, allowing actors to discover and exploit opportunities for goal attainment and goal adjustment as they arise. In line with much of the logistics literature, collaborative planning is suggested as central to establishing and maintaining effective relationships within and across organizations. The ultimate goal of collaborative planning has to be the fair distribution of uncertainty across the different partners involved. From this premise, a model of collaborative planning is developed and evidence for its validity provided from two empirical studies; one on interdepartmental planning and one on planning in supply chains. Based on the results, it is argued that organizational structures supporting cross-departmental cooperation and both perceived interdependence and perceived autonomy foster collaborative planning. Finally, a case study is presented to illustrate the use of cause maps for identifying belief systems as crucial determinants of collaborative planning.

In Chapter 6, the design of automated systems is discussed by elaborating on the issue of control and accountability in such systems. Regardless of a system's degree of automation, it is humans who are responsible for its functioning. The basic tenet of human-centred automation – that human operators need to be in control of technical systems – is derived from this responsibility, referring back also to the necessity of enabling coping with uncertainties. KOMPASS is presented as an example of methods aimed at supporting managers and engineers in designing systems that are controllable by the human operator especially in those conditions where all other safeguards fail. In view of human fallibility in general, as well as loss of control evidenced in accidents in particular, doubts are expressed about how achievable the goal of human control over technology really is. Reasons for lack of controllability can be found in the normative assumptions of those developing and implementing technology which may support a self-fulfilling prophecy of turning humans into risk factors. However, the ever increasing complexity of systems also has to be acknowledged as a limiting factor. It is therefore suggested to found system design on the premise of partial non-controllability of technology. This approach could help human operators to deal with system opaqueness and uncertainty better by providing systematic information on the limits of control and thereby also relieving them of some of their responsibility. At the same time, this approach would force system designers, the organizations operating the systems, and regulatory institutions to take on responsibility for the use of technical systems whose complexity can no longer be mastered entirely. Consequences for making decisions on system automation by regulators and companies are discussed within the larger realm of establishing a new politics of uncertainty on a societal level, which would be based on deliberately giving up the pretence of being in control. Finally, some comments are made regarding the basic relationship between technology and organization and the limitations on systematically establishing particular use patterns through technology design.

In Chapter 7, uncertainty management is examined from the perspective of ensuring safety as a crucial organizational outcome, with a particulate focus on companies whose work processes contain high environmental and health risks. Safety has long been managed mainly by attempting to minimize uncertainties following the basic understanding of safety as minimal acceptable risk. However, minimizing uncertainties implies losing the organizational flexibility needed for resilient responses to disturbances. Therefore, newer approaches to safety management aim to combine different ways of handling uncertainties into strategies that improve the balance between stability and flexibility in line with the notion of loose coupling. Safety culture is seen as important for achieving loose coupling because culture can serve as a "soft" mechanism for coordination and integration in a formally decentralized organization. As there is much reluctance to adopt the coping with uncertainties approach in high-risk organizations due to the basic assumption that individual autonomy furthers errors, the empirical evidence on the relationship between safety and autonomy is discussed and uncertainty postulated as an important moderator in that relationship. This provides further justification for approaching safety management and related design questions within a framework of uncertainty management. Following this thread, a new instrument and assessment procedure for evaluating and designing safety management and safety culture is proposed. The procedure emphasizes the importance of assessing the fit between uncertainties and their management and of changing organization design to improve the fit if needed. Change efforts should mostly worry about culture in terms of current belief systems getting in the way of establishing better organizational practices. However, in order to advance loose coupling, attention also needs to be paid to furthering culture as a mechanism for integration and coordination.

Chapter 8 switches perspective and discusses uncertainties from the viewpoint of the individual employee. Firstly, job design literature is reviewed with respect to the demands on handling uncertainties in everyday work processes. Job autonomy, meaning involvement in decisions on goals and ways of achieving them, and job control, meaning the actual means to achieve goals in a particular work situation, are identified as crucial prerequisites for individual management of operational uncertainty. Secondly, uncertainties in the employment relationship through increasing flexibility of working arrangements are discussed, demanding individual acceptance of less job security and more effort to stay employable. In particular, consequences of flexibility are described in terms of changing psychological contracts, new forms of more flexible and independent careers, and increasing demands on employability. In order to manage uncertainties as part of human resource management better, forecasts of skill requirements based on the strategic positioning of the company are necessary; employer and employee expectations, contributions, and inducements within the framework of psychological contracts need to be monitored and adjusted continuously; employment relationships have to cater for individual resources and needs regarding uncertainty management, as expressed, for instance, in traditional organization-bound versus independent, self-reliant career orientations.

In the concluding chapter, some final observations are made across the different domains treated in the book. A significant benefit of the analytic framework and the generic process proposed for making decisions on the management of uncertainty is seen in supporting a concerted approach across different areas of organizational functioning. An example from an extensive analysis in a railway company is given to substantiate this claim. Finally, some research questions are identified concerning the different domains and the general approach to managing uncertainty. Furthermore, practice implications are summarized in view of a future that will bring even more uncertainties.

2

Defining and Identifying Uncertainties in Organizations

Uncertainty is about not knowing for sure. In organizations, this may concern daily business and entail questions such as: Will the material be delivered on time for next week's express orders? How many doctors should be scheduled to cover the demand in the shock trauma centre over the holiday season? Will the newly hired product development team cooperate well? On a more strategic level, uncertainty is encountered in decisions such as: Should we spend more on research and development? Which qualifications do our employees need to improve customer orientation? Should we increase the level of standardization in our production processes? Uncertainty is also involved in many personal decisions: Should I follow the production schedule even though it is suboptimal? Will the new job offered to me allow me to develop my career better?

In trying to handle uncertainties, one needs to know first of all which kinds of uncertainties one is currently facing. Also, it is important to address future uncertainties, which adds to the complexity; one is now uncertain about uncertainties. In order to carry out such an analysis, the very first step is to define what one is looking for, that is, to define uncertainty.

In the following, a general framework will be developed that distinguishes causes, sources, contents, and consequences of uncertainty, as well as different actors involved in the management of uncertainty. It centres on the uncertainties' effects for controlling action towards achieving certain outcomes. The framework's elements can be applied to analyze conditions of managing uncertainty from a strategic and an operational perspective and examples for both will be provided. Moreover, the framework allows the adoption of a rationalistic, objective stance on managing uncertainty, but also the consideration of the impact of individual and collective enactment and sensemaking. The overall aim of the analytic framework is to help the decision-makers in organizations to sketch an uncertainty landscape that supports informed decisions on managing these uncertainties so as to achieve desired outcomes. At the end of the chapter, an example of an analysis carried out within the suggested framework will be presented.

2.1 Definitions of Uncertainty

Turning to organization science for a definition of uncertainty, one quickly finds that there is no easy answer. As the smallest common denominator, one can take the first sentence of this chapter "uncertainty is about not knowing for sure". To state this definition somewhat more scientifically: uncertainty is the absence of information and more specifically, the difference between the amount of information required to perform a task and the amount of information already possessed by the organization (Galbraith 1973). Additionally, the ambiguity of existing information in terms of the multiplicity of meanings that can be imposed on a situation, has been considered an element of uncertainty, sometimes also referred to as equivocality (Weick 1979; Daft and Lengel 1984; Leifer and Mills 1996).

In order to make these definitions operationally useful, one needs to specify further what information or lack thereof is relevant for the functioning of an organization. Given the multitude of attempts to do just that in academic literature, it becomes obvious that rather than looking for *the* definition of uncertainty, it is important to choose a definition that fits the nature of the organization and the nature of the question studied (Kreiser and Marino 2002). The following review of approaches to defining uncertainty is intended to help researchers and practitioners who are interested in studying uncertainties in organizations to develop a definition of uncertainty that is both sound and fits their needs. For that purpose, different perspectives on uncertainty as found in the scientific literature, are discussed and examples – mostly from classic studies in the field – given on how researchers have selected the perspective and corresponding definition most useful for a particular analysis of uncertainties.

The overall framework suggested for the analysis of uncertainties is presented in Figure 2.1. Its rationale is developed in the following sections and an example for its use in an actual analysis of uncertainties contained in train operations is given at the end of the chapter.

2.1.1 Content of Uncertainty

Frequently, uncertainty is described as a characteristic of decisions that individual decision-makers in organizations have to take (e.g., Beckert 1999; Eisenhardt 1989; Lipshitz and Strauss 1997; Palmer and Wiseman 1999; Sitkin and Pablo 1992). Decisions are understood as a choice being made between alternatives differing in the usefulness and likelihood of their outcomes, that is in their expected utility. From this perspective, uncertainty may concern the probability of an event (state uncertainty), a lack of information about the outcomes of an event and the underlying cause-effect relationships (effect uncertainty), or a lack of information about response options and their likely consequences (response uncertainty) (Milliken 1987). In Milliken's original work, the three contents of uncertainty are described in terms of environmental uncertainty, so for instance, effect uncertainty is understood only as uncertainty about impacts of environmental events on the organization. Here the distinction is used more broadly to define the different contents of what the decision-maker is uncertain about, that is, any state,

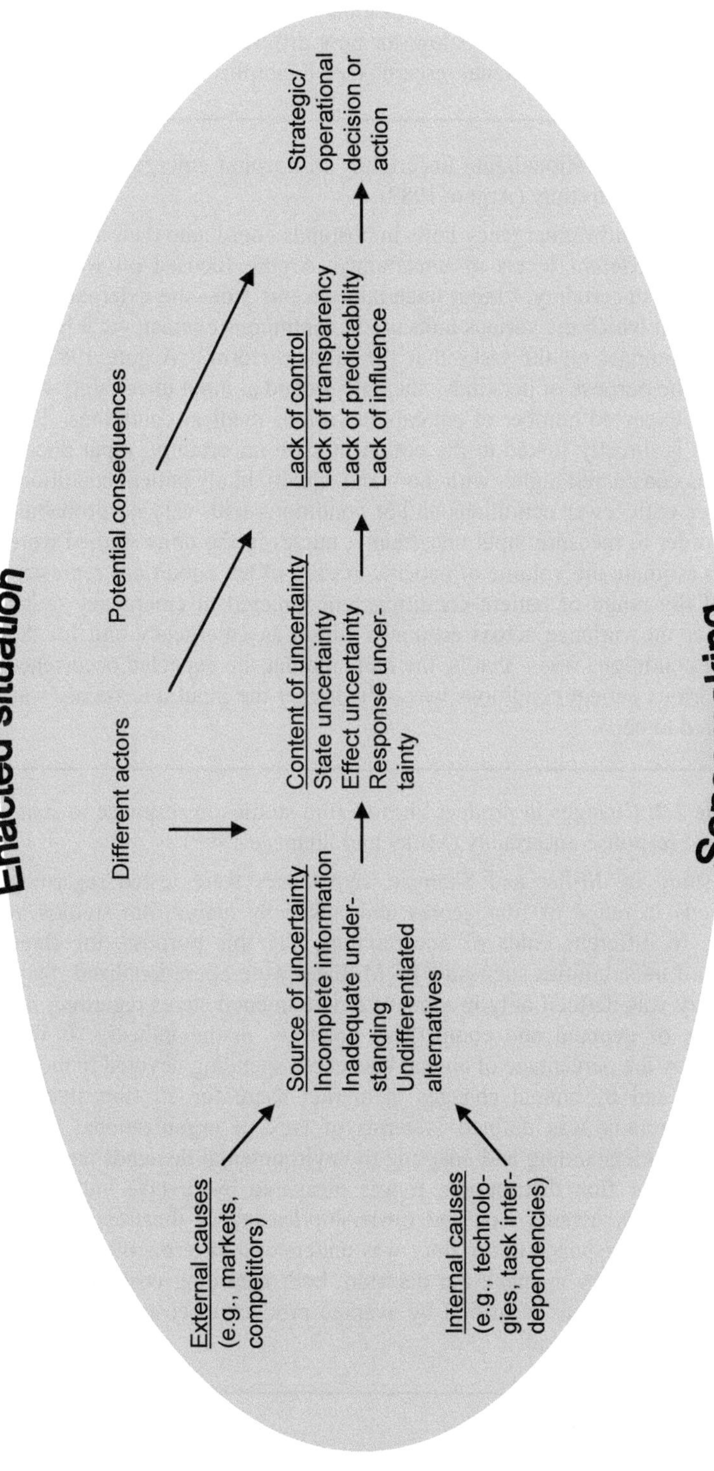

Figure 2.1. Framework for uncertainty analysis

any effect, or any response in the organization or in the environment. Two very different examples are presented below for how different contents of uncertainty have been defined in organizational research (see Examples 2.1 and 2.2).

Example 2.1. Operationalizing uncertainty in hospital emergency units in terms of state uncertainty (Argote 1982)

In her study on how emergency units in hospitals coordinate their actions in the face of different levels of uncertainty, Argote focused on what she termed input uncertainty. "Input uncertainty stems from the external environment with which the various units are in continuous contact, yet it has an immediate impact on the tasks that the units perform" (Argote 1982, p. 422). For the purpose of her study, she only looked at input uncertainty with respect to expected number of patients in various medical conditions. This definition is directly linked to the notion of state uncertainty. Input uncertainty was considered higher with more and equally likely patient conditions and lower with fewer conditions and/or conditions with varying probabilities. In order to measure input uncertainty, nurses in the units studied were asked to estimate the volume of patients in each of ten conditions representative of the range of patient conditions encountered in emergency units. The lower the variance across estimates within an emergency unit for the different conditions was – that is, the more similar the expected occurrence of the various patient conditions was – the higher the input uncertainty was considered to be.

Example 2.2. Changes in product lines of film studios in response to state, effect, and response uncertainty (Miller and Shamsie 1999)

In the study by Miller and Shamsie, hypotheses were tested regarding adaptations in range of film genres undertaken by major film studios in response to different kinds of uncertainties. For this purpose, the three contents of uncertainties suggested by Milliken were operationalized. State uncertainty was defined only in view of environmental states regarding the robustness of demand and competitive volatility in the industry. It was measured by the percentage of annual household spending devoted to movie attendance and by annual changes in market share for all film studios. Effect uncertainty was defined in terms of lack of organizational skills needed for understanding and adapting to environmental demands and lack of control over film distribution. It was measured by reverse indicators, namely academy awards won and ownership/leasing of theatres per film studio. Finally, response uncertainty was understood in terms of risks perceived by managers in making a decision, both regarding costs and their own reputation. It was measured by average production cost per film and tenure of production head.

2.1.2 Sources of Uncertainty

Within the general decision-making framework employed here, uncertainty can arise due to incomplete information, inadequate understanding of available information, and undifferentiated, that is equally attractive or unattractive, alternatives (Lipshitz and Strauss 1997). The latter may also be related to uncertainties regarding the goals to be achieved by making a certain decision. While incomplete information is a factor that can be determined objectively, both undifferentiated alternatives and inadequate understanding are sources of uncertainty that imply an interaction between characteristics of the decision to be taken, the environment in which the decision is embedded, and the decision-maker him- or herself (Sitkin and Pablo 1992). Because of these differences, sometimes only incomplete information is defined as uncertainty, whereas the other two elements are subsumed under a separate category ambiguity.

The attractiveness of alternatives is linked to the goals, values, needs, and attitudes of the decision-maker, as well as the expected utilities for the organization in whose interest the decision-maker is supposed to act. Inadequate understanding may be a consequence of too little, but also of too much, information. Moreover, it may be caused by information that is difficult to interpret because several, potentially even conflicting, meanings exist (Weick 1979; Leifer and Mills 1996). The decision-maker's competence for handling large amounts of ambiguous information will strongly influence the degree of uncertainty experienced in those cases. Given the importance of decision-makers themselves as defining factor for uncertainties, it has been frequently argued that measures of uncertainty should focus on the uncertainty as perceived by decision-makers instead of on objective accounts of uncertainty in the decision-making situation (see Example 2.3 for a prominent study that took this approach). We will return to this important point later.

Example 2.3. Exploring contingencies between uncertainty and organizational design (Lawrence and Lorsch 1967)

In their seminal study on contingencies between uncertainty and organizational design, Lawrence and Lorsch measured uncertainty by means of top executives' perception of clarity of information, uncertainty of causal relationships, and time span of definite feedback, thereby touching upon two of the mentioned sources of uncertainty, that is incomplete information and inadequate understanding of available information. The three questions asked to measure the three components of uncertainty were the following (Lawrence and Lorsch 1967, p. 248ff):

a) Please circle the point on the scale provided which most nearly describes the degree to which present job reqirements in each functional department are clearly stated or known in your company for:

Research	Job requirements	1 2 3 4 5 6 7	Job requirements
Manufacturing	are very clear in		are not at all clear
Marketing	most instances		in most instances

b) Please circle the point on the scale provided which most nearly describes the degree of difficulty each functional department has in accomplishing its assigned job, given the limitation of the technical and economic resources which are available to it.

Degree of difficulty in:

Developing a product which can be manufactured and sold profitably	1 2 3 4 5 6 7
Manufacturing economically a product which can be designed and sold.	Little difficulty — Extremely difficult
Selling a product which can be developed and manufactured economically.	

c) Please check the alternative which most nearly describes the typical length of time involved before feedback is available to each functional area concerning the success of its job performance.

Research Department	(1) one day
Manufacturing Department	(2) one week
Marketing Department	(3) one month
	(4) six months
	(5) one year
	(6) three years or more

2.1.3 Causes of Uncertainty

Beyond determining the different sources of uncertainty, that is incomplete information, inadequate understanding of information, and undifferentiated alternatives, one can study their *causes*. One might be interested in personal causes of uncertainty, for instance, why a person is more or less able to handle ambiguous information due to different cognitive abilities or personality characteristics like control orientation or emotional stability. Especially different risk preferences and thereby different attitudes to uncertainty, have received much attention in the literature (*e.g.,* MacCrimmon and Wehrung 1990; Sitkin and Pablo 1992). However, as the focus of this book is on organizational requirements for managing uncertainties, we will discuss these personal causes mostly in passing in the following chapters.

Situational causes can be found in the external environment of the organization, *e.g.,* suppliers not providing information on anticipated late deliveries, or in the organization's internal processes, *e.g.,* unforeseen complications in heart surgery. In academic literature, environmental uncertainties have been treated much more prominently (*e.g.,* Bourgeois 1985; Eisenhardt and Martin 2000; Kreiser and Marino 2002). Ever since the shift from viewing organizations as closed systems acting more or less independently of their environment to acknowledging their nature as open systems co-existing with and depending on their environment, environmental uncertainties have been a core concern (Thompson 1967). Pfeffer and Salancik's (1978) eminent conceptualization of organizations' dependence on their environment focuses on the dependence on resources from external partners with varying degrees of power as the core cause of uncertainties. In their view, uncertainty – defined as "the degree to which future states of the world cannot be anticipated and accurately predicted" (p. 67) – is only problematic when "it involves an element of critical organizational interdependence" (p. 68). In resource dependence theory, only those uncertainties need to be managed that arise due to the criticality of a particular resource for the organization's survival, the scarcity of that resource, and the level of competition between organizations for that resource. These characteristics of the environment are crucial in determining the influence an organization can exert to obtain necessary resources and the different strategies available to control resource access. Uncertainty therefore is understood in terms of lack of control, and not as lack of information as such. The consequences of this distinction will be taken up again later in this chapter.

Internal uncertainties have been described less succinctly, usually listing different functions and operations in the organization that can cause variability and unpredictability of work tasks. Therefore, internal uncertainties are often summarized under the term task uncertainty (Van de Ven *et al.* 1976). In Perrow's (1967) prominent analytic framework for comparing organizations, variability and lack of knowledge concerning raw materials and technology are mentioned as key factors. Slocum and Sims (1980) distinguish between task uncertainty, related to lack of knowledge about how to produce the desired outcome, and workflow uncertainty, which concerns lack of information about inputs and outputs for a particular work station. In the literature on sociotechnical design, internal (but potentially also external) uncertainties are frequently termed variances, defined as "any unprogrammed event", and particularly important ones, which critically affect outcomes, are termed key variances (Cherns 1976, p. 787). Examples given are often lists of various possible events in an organization, such as insufficient quality of raw materials or machine failures.

A related concept frequently discussed in the sociotechnical literature is that of task interdependence (Thompson 1967; Van de Ven *et al.* 1976). Individual tasks are linked through technical and organizational design in ways both creating uncertainties and allowing for particular ways of handling uncertainties. Usually three types of task interdependence are distinguished:

Pooled interdependence is present when system performance is an additive function of individual performance. The performance of other members of the system may affect the work of the individual members, but only indirectly through parallel contributions to a superordinate goal. Coordination in this case is usually achieved via centrally determined work programmes which every individual has to follow independently and which assure that subtasks serve the superordinate goal. An example would be a service organization, such as an insurance company, where individual employees are responsible for all the concerns of a particular group of customers.

Sequential interdependence is a unidirectional workflow arrangement, where individual performance depends on the proper fulfilment of prior subtasks. Synchronization is needed based on centrally determined programmes and plans that spell out the exact content and temporal requirements of subtask fulfilment. The classic example of this kind of interdependence is the assembly line.

Reciprocal interdependence implies that information and results of work activities have to be exchanged between team members continuously. Coordination is mainly achieved via direct communication, be it in the form of personal directives, or multilateral flow of communication between the individuals involved in self-regulated task performance. Project teams of any kind are examples of this type of interdependence.

Each type of task interdependence involves particular causes of uncertainty for the individual actors. For instance, in a situation of pooled interdependence, uncertainty is mainly created by inappropriate programmes used to coordinate individual tasks, as well as by problems in the fulfilment of individual tasks such as machine breakdown. Sequential interdependence implies that uncertainties are handed down along the process and, if not handled adequately, create a series of problems throughout the process. Reciprocal interdependence involves multiple parallel sources of uncertainties, such as misunderstandings about task requirements, changes in individual plans for task fulfilment, or inadequate consideration of interfaces in project specifications. In Example 2.4, the operationalization of uncertainty used in a study on different ways of handling task interdependencies is described.

Example 2.4. Testing contingencies between task uncertainty, task interdependence and coordination modes (Van de Ven *et al.* 1976)

Hypotheses regarding the relationships between task uncertainty, task interdependence, and coordination modes were tested by means of questionnaire data collected in a large state employment security agency. Task uncertainty defined as the "difficulty and variability of the work that is undertaken by an organizational unit" (Van de Ven *et al.* 1976, p. 333) was measured by eight items measuring perceived uncertainty (*e.g.,* During the course of your work, how often do you come across specific but difficult problems that you don´t know how to solve, and you have to take some time to think through by yourself or with others, before you can take any action? How much variety in cases, claims, clients, or things do you generally encounter in your normal working day?) and by a standardized classification of tasks. This classification was based on lists provided by each respondent of all tasks performed on a normal day and the percent of time spent on each task. The tasks were classified by the researchers according to difficulty and variability and individual scores were calculated for the predominant task performed by each respondent. The individual scores were then averaged per organizational unit and correlated with the perceived uncertainty scores as a validity check.

Additional uncertainties, irrespective of the particular form of interdependence, may be created if the individuals involved in the different work arrangements are not provided with the adequate means to manage task interdependencies (Crowston 1997; Tushman and Nadler 1978). For instance, in pooled interdependence this might imply that there are insufficient opportunities to adapt central work programmes to changing circumstances. In reciprocally interdependent team work, problems may arise when mutual adjustment is impeded by lack of direct communication. This particular issue has been studied extensively in distributed project teams where much of the communication has to happen via electronic means (*e.g.*, Hinds and Mortensen 2005; Kraut *et al.* 1999).

2.1.4 Organizational Actors Affected by Uncertainty

Two types of actors can be distinguished in the literature on handling uncertainties in organizations: high-level executives in their role as strategic decision-makers and operative personnel making decisions in the course of day-to-day work processes. The basic definitions of uncertainty are usually similar, their operational meaning is, however, very different for these different actors. Pfeffer and Salancik (1978) and Wall and colleagues (2002) for instance, both describe uncertainty as lack of knowledge and lack of understanding of cause and effect relationships. However, Pfeffer and Salancik further specify uncertainty in terms of coping with critical organizational interdependence on a strategic level, while Wall and colleagues define what they call operational uncertainty further in terms of "lack of knowledge about production requirements, of when problems will be met, and how best to deal with them" (Wall *et al.* 2002, p.159)

In order to understand how uncertainties affect decision-making in organizations, one can easily imagine that several kinds of actors should be considered: Is it top management having to make decisions such as which company to buy, which products and markets to develop, or which strategic alliances to build? Is it middle management having to mediate between strategic decisions and operations, for instance, by setting up production facilities in a way that allows for faster changes between products? Is it shopfloor personnel having to handle variations in production schedules, machine breakdown, and customer demands? Is it people in line functions, in core business units, or people in staff functions? Relevant sources and effects of uncertainties will differ substantially depending on the group of actors and types of decisions and actions required.

There are only few studies which have systematically addressed different groups of actors in relation to the management of uncertainty. One classic example is again the study by Lawrence and Lorsch (1967), which had the explicit aim of comparing contingencies between uncertainty and organizational design for different departments in organizations, specifically R&D, marketing, and manufacturing. However, the actual measure of perceived uncertainty in these different departments was derived from the account of the top executive in each company alone (see Example 2.3).

2.1.5 Potential Consequences of Uncertainty

A final perspective concerns the potential consequences of uncertainty. Thompson (1967) has described task uncertainty as the inability to act deterministically. With that same focus on action, Lipshitz and Strauss (1997, p. 150) defined uncertainty as “a sense of doubt that blocks or delays action”. In Pfeffer and Salancik’s (1978) resource dependence theory, only those uncertainties are considered relevant that affect critical organizational relationships. The focus on consequences of uncertainties may help to identify those uncertainties that have to be managed most urgently. This perspective can thereby act as a useful filter for defining and subsequently analyzing uncertainties in an organization.

Lipshitz and Strauss (1997) also point out that doubts experienced in decision-making depend on the chosen decision-making model, which in turn highlights certain consequences and disregards others. For instance, applying an expected utility model focuses attention on doubts regarding possible alternatives, their outcomes, and the attractiveness of these outcomes in relation to certain goals. Considered consequences therefore are directly linked to capitalizing on positive outcomes, *e.g.*, raising revenue by buying a particular firm. A role compliance model, on the other hand, implies handling doubts about situational requirements in light of an assigned role. Possible consequences of fulfilling or not fulfilling that role are crucial considerations, *e.g.*, will I still be considered a proactive manager if I abstain from changing the firm’s production portfolio? The uncertainties selected for analysis and management vary accordingly. This again points to the importance of distinguishing objective and subjective approaches to measuring uncertainty, which will be discussed further in a later section of this chapter.

2.2 Defining Uncertainty within a Control Framework

In line with the suggestion that one should choose definitions of uncertainty that fit the nature of the organization and the nature of the question studied, different perspectives on uncertainty have been presented that should be taken into consideration when making this choice. Starting from the perspective of an individual decision-maker having to choose between different alternatives in view of achieving particular goals that themselves may not be fully clear, three contents of uncertainty were distinguished: uncertainties concerning states, effects, and responses. Subsequently, three sources of uncertainties were discussed: incomplete information, inadequate understanding of available, but ambiguous information, and undifferentiated alternatives. Looking into situational causes of uncertainties, environmental and internal uncertainties were distinguished and resource dependence was discussed as a prominent contributor to environmental uncertainties. Additionally, task interdependencies were discussed as an important cause of internal uncertainties. Finally, it was pointed out that uncertainties may impinge on different kinds of actors and have different kinds of consequences for these actors.

From the examples provided along with the discussion on different definitions of uncertainty, it can easily be seen that many different ways of defining uncertainties are possible, leaving the choice to the researcher or practitioner. As this book ultimately wants to help decision-makers in organizations in not merely analyzing, but also shaping the ways they manage uncertainties, Pfeffer and Salancik’s conceptualization of uncertainty in terms of lack of control is suggested as the basic framework. That is, uncertainties should always be studied from the perspective of control over decisions and the different actions required for the achievement of organizational objectives. "The fact that we expect all organizations to seek the same state – self-control – does not mean that we expect all of them to attain it in the same way, with identical design, structures, or behavior. It is essential that we find universals, but equally essential to find patterns in variations" (Thompson 1967, p. 161).

In the broadest sense, control is usually described as exercising influence or power in order to reach certain goals, often also with the implication of influence or power over other people. From a system theory perspective, control entails more specifically regulation via feedback (*e.g.,* Green and Welsh 1988). In order for influence to be exerted effectively, there has to be sufficient predictability and understanding or transparency of the situation (Brehmer 1992; Sutton and Kahn 1987), that is: little uncertainty. Turning this argument around, uncertainty reduces effective control because it reduces transparency and predictability. Additionally, there may even be uncertainty as to which means of influence there are and what effects they have on what outcomes, which is implied in the notion of response uncertainty introduced earlier. While lack of transparency and predictability can be fully described in terms of uncertainty, lack of influence is predominantly determined by other factors, such as the distribution of power within and across organizations and the competence level of the actors. However, as transparency and predictability are necessary prerequisites for using existing means of influence, the degree and types of influence available in a particular situation are directly affected by uncertainty. Influence also involves the active handling of uncertainties, for instance, by reducing uncertainties or shifting uncertainties to less powerful others (Marris 1996; Pfeffer and Salancik 1978). Therefore, uncertainty and influence are reciprocally related to each other. Consequently, influence is included in the definitional framework even though it cannot be fully specified by the factors contained in the framework.

In Figure 2.1, the different facets to be considered when defining uncertainties for a particular purpose are presented. The arrows in Figure 2.1 are not be taken as signalling cause-effect or antecedent-consequence relationships, but they indicate rather a possible ordering of facets in a particular analysis in line with "mapping sentences" used in facet theory (e.g., Guttman and Greenbaum 1998, Shye 1998). For instance, an analysis may concern externally caused inadequate understanding regarding response options for employees in a sales department. A concrete example within such an analysis could be customers who have the habit of vaguely announcing the possibility of orders, which each time leaves the sales agents wondering whether or not they should raise their forecast. The facets of sense-making and enactment have not been discussed so far because they concern the issue of employing objective versus subjective accounts of uncertainty when identifying and managing uncertainties. This will be covered in the next section.

2.3 Identifying Uncertainties in Organizations

The framework presented in Figure 2.1 is meant to guide, but not determine uncertainty analysis. The overall aim of the analytic framework is to help decision-makers in organizations to sketch an uncertainty landscape that supports informed decisions on managing these uncertainties so as to achieve desired outcomes. Depending on the type of organization and the particular problems to be dealt with in that organization, different elements of the framework will be of more or less importance. The examples in the previous section illustrated how the particular focus of an analysis will determine the facets of uncertainty to be studied. Before another more complete example of the application of the framework is given, the fundamental question of objective versus subjective approaches to identifying uncertainties is discussed.

2.3.1 Objective versus Subjective Accounts of Uncertainty

In the literature on assessing and managing uncertainties, two prominent debates can be found: (1) the significance of objective versus subjective accounts of uncertainty; (2) whether organizations and the decision-makers within them merely react and adapt to environmental uncertainties or actively shape their environments as well (Jauch and Kraft 1986).

Regarding the first debate, the focus on decision-makers in organizations and their having to face uncertainties has fuelled arguments against objective accounts of uncertainty, in favour of perceptions of uncertainty by the actors themselves as adequate measures. Some of the facets included in Figure 2.1 in fact, imply measurement of perceptions, such as inadequate understanding of information, which can only be assessed for a particular actor. Contributing factors for inadequate understanding, such as ambiguous information, again need to be identified with reference to the actors affected. Many landmark studies on contingencies between uncertainty and organization design therefore, only used perceptual measures of uncertainty (see Example 2.3). Also, perceptual accounts of uncertainty have been used to further the understanding of the concept of uncertainty itself. Duncan (1972), for instance, used the two dimensions "simple-complex" (related to the number of factors to be taken into account in a decision) and "static-dynamic" (related to the stability of the factors over time) to study decision-making in organizations, and found that the static-dynamic dimension was more important for the experienced level of uncertainty.

However, other authors have argued that both objective and subjective accounts of uncertainty need to be assessed and that in fact, the co-alignment of perceptions and objective realities is crucial for good performance (Bourgeois 1985). The validity of this finding can be questioned in light of the second debate.

This second debate delves into basic beliefs by researchers and practitioners alike about the relationship between organizations and their environment. One of Pfeffer and Salancik's (1978) main intentions in developing resource dependence theory was to show both organizations' links with their environments and how they can actively shape these links. In doing so, organizations do not react to or act upon given realities, but they enact their environments through processes of selective attention and interpretation. Or as Weick (1979, p. 130) puts it "Enactment is the only process where the organism directly engages an external environment. All processes subsequent to enactment work on edited raw materials […]." Therefore, "the question of what the environment is, is meaningless without regard to the focal organization which enacts it, or more precisely, the individuals who enact it in planning the activities of the organization" (Pfeffer and Salancik 1978, p. 73). Enacting situations is highly intertwined with processes of sensemaking (Weick 1995), that is, literally making sense out of things happening through placing them in a particular framework and deriving their meaning from that. Environments are socially constructed on the basis of actors' subjective theories and preconceptions which are used to make sense of any ambiguous information. Any meaning that is derived from this sensemaking process subsequently shapes actors' behaviour. It is through the cyclical sensemaking and enactment processes that actors can either improve their own conditions or, in contrast, can become the authors of their own problems (Weick 1979). Self-fulfilling prophecies are an extreme case of sensemaking, because starting from specific assumptions about the world, people act in ways that will make these assumptions come true. "People create and find what they expect to find" (Weick 1995, p. 35).

Sensemaking is particularly powerful under conditions of high uncertainty when incomplete or ambiguous information requires assumption-based reasoning and interpretation in order to develop a basis for action (Lipshitz and Strauss 1997). Weick points out that uncertainty stemming from incomplete information should be clearly distinguished from ambiguity based on several possible meanings of information. Ignorance and confusion are not the same thing and require different sorts of action, as Daft has studied (Daft and Macintosh 1981; Daft and Lengel 1984): To remove ignorance, more information is needed; to reduce confusion, richer information is needed that provides more cues for sensemaking, for instance, rather a face-to-face conversation than an electronic message.

Questioning the objective and deterministic nature of uncertainties even further, Weitz and Shenhav 2000, p. 246) have argued that notions of uncertainty in the internal and external organizational environment are socially constructed by professionals dealing with management of organizations in order to legitimate the rise of management as such and "to justify […] managerial and organizational methods of control". They tested their claim with respect to the rise of scientific management at the end of the 19th and beginning of the 20th century using data on the discourse on technical and organizational uncertainty in two prominent American mechanical engineering journals and linking it to membership numbers in the American Society of Mechanical Engineers and to labour unrest. Given that discourse on technical uncertainty, society membership, and labour unrest all added significantly to explaining the amount of organizational uncertainty discourse, Weitz and Shenhav postulated that engineers used their success in handling technical uncertainties to expand their professional domain to include the reduction and elimination of organizational uncertainties as well. The pervasiveness of assembly lines to this present day and the substantial efforts by many companies to shape their environments accordingly provide powerful support for this claim. Barley and Kunda (1992) argued more generally that just how uncertainties are understood and which form of uncertainty management is chosen depends on the prominence of normative versus rational ideologies; that is, ideologies that build on normative control such as culture, versus ideologies that build on rational control, such as Taylor`s scientific management. The prevalence of these ideologies is, in turn, influenced by changes in labour relations and expansions/contractions of the economy. Prospering of normative ideologies seems to coincide with economically difficult times and little labour unrest, while rational ideologies tend to parallel economic expansion and confrontational labour relations.

Uncertainty analysis should, at best, include both perception-based measures and objective indicators. The combination of both approaches helps to account for individual and collective processes of enactment and sensemaking, while at the same time allowing the identification of and compensation for selective processes of information uptake and interpretation.

2.3.2 Uncertainty Analysis: An Example from Railway Operations

The framework presented in Figure 2.1 was used to analyze the uncertainties contained in a particular safety-relevant task in a railway company, which had been related to several incidents and accidents in the recent past (see Figure 2.3 for a summary of the analysis). The task concerns moving a train from a track used only for shunting onto an occupied track from which it is to depart as a passenger train (see *required decision/action* in Figure 2.3). This operation is carried out by the train driver without the assistance of shunters and is guided by dwarf signals. Usually, signals for the train to move onto a track are given when the track sector concerned is free. A special situation, which increases the risk of this operation, is moving the train onto a track sector that is still occupied by another train. This is sometimes done when traffic is particularly dense, for instance, when a train is still on the track but is to depart at the same time as or very soon after the shunted train is supposed to get there (see *external cause* in Figure 2.3).

The task requires cooperation of two actors; the train driver – who also acts as shunting supervisor because no shunter is on the train with him or her – and the signaller (see *actors* in Figure 2.3). The shunting supervisor's responsibility is to determine and lead shunting movements, which includes determining the track to be used, informing all persons affected by the planned train movement, checking several operational requirements (*e.g.,* release of brakes), and requesting a go signal for the chosen track. Also, the shunting supervisor is to be informed by the signaller if there are unexpected obstacles on the chosen track, for example another train on the same track sector. Shunting operations are signalled by means of dwarf signals. These can take on several states, with not all of them having singular meanings (see *internal cause* in Figure 2.3). In particular, one signal state frequently used for shunting in stations "go with caution" implies three possible meanings: expect an obstacle, expect a stop signal at the next sector, or there is no further dwarf signal (see Figure 2.2). Depending on the meaning assigned to the signal by the train driver, for instance, based on his or her expectations linked to prior experience with the particular situation, he or she will drive with the maximum speed allowed of 40 km/h or drive more slowly, in order to be able to stop quickly (see *sources of uncertainty* in Figure 2.3). From their instruments, signallers are not able to ascertain the actual speed of the shunted train; they only have an

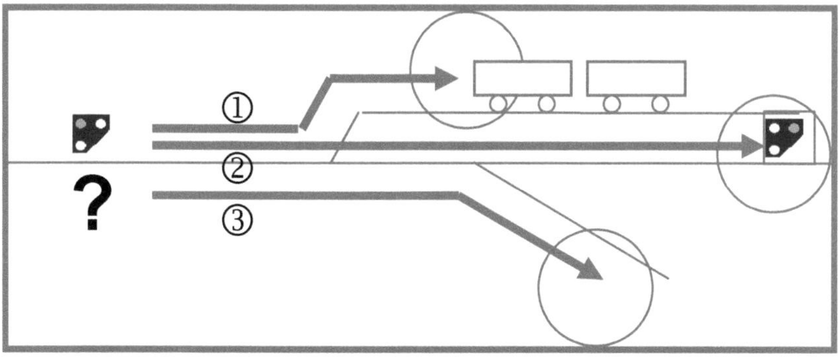

Figure 2.2. The three possible meanings of the dwarf signal "Go with caution"

indication of the train moving across the different track sectors (see *sources of uncertainty* in Figure 2.3). Therefore, they cannot predict the exact arrival time of the shunted train on the occupied track and cannot plan more specifically as to whether the departing train on that track will be gone by the time the shunted train gets there (see *content of uncertainty* and *lack of control* in Figure 2.3).

The standard operating procedures have recently been changed regarding the role of shunting supervisors. For unaccompanied trains the signaller used to be the shunting supervisor; with the new rules the train driver is assigned this role. Several incidents have happened where, among other issues, it was obvious that this new role assignment had not been implemented, but rather signaller and train driver still behaved in accordance with the old role distribution (see *enacted situation* in Figure 2.3). This implied that the train driver did not select the track he or she was to move to. Also, the train driver was not told explicitly that the track to move to was occupied (see *sources of uncertainty* in Figure 2.3). Instead, the signallers' implicit assumption was that the train drivers would be ready to meet an obstacle because this is one of the meanings of the employed dwarf signal state "go with caution". Also, the maximum speed of 40 km/h should only be used when the track the train is moving to can be seen by the train driver, which was not the case in the incidents analyzed. The train drivers, on the other hand, took the dwarf signal "go with caution" in combination with the absent communication from the signallers concerning obstacles to mean that they were about to move to a free track (see *sensemaking* in Figure 2.3). They therefore chose the maximum speed of 40 km/h, leading in a few cases to critical situations or even a collision with the train already on the track (see *potential consequences* in Figure 2.3). Overall, the handling of this situation by signaller and train driver involved high state and high response uncertainty (see *contents of uncertainty* in Figure 2.3) and led to a lack of transparency and predictability for the train driver and a lack of predictability and influence for the signaller (see *lack of control* in Figure 2.3).

The analysis shows that either an enforcement of the new role assignment or at least more direct communication between signaller and train driver would be required to reduce the uncertainty for the train driver. As the signaller's work load is often too high to expect him or her to spend extra cognitive resources and time

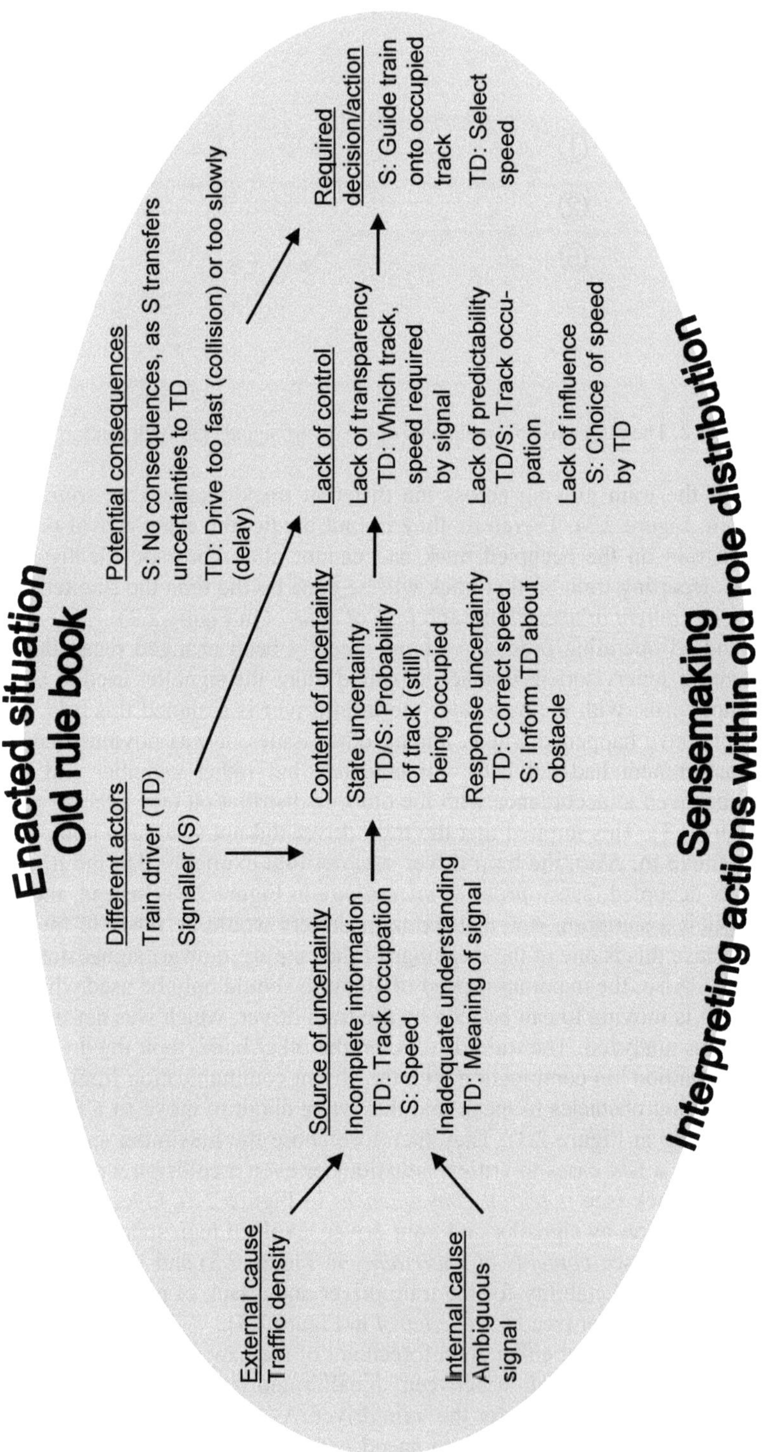

Figure 2.3. Analysis of uncertainties involved in shunting a train onto an occupied track

on communication, another option would be to generally lower the speed the train is allowed to move at upon receiving the dwarf signal and so not reducing uncertainty, but reducing potential negative consequences of the uncertainty. As using occupied tracks is often related to particularly high traffic density and time pressure, this decision could be counterproductive because it might create even more time pressure. These brief considerations already illustrate that the design of a presumably straight-forward operational task may involve strategic decision-making as well, in this case, decisions on production capacity and interaction between traffic control and train control. A more detailed analysis of different options for managing uncertainties better for this task and the concrete steps taken by the railway company to handle this situation will be described at the end of the next chapter.

2.4 Chapter Summary

In this chapter different perspectives on defining uncertainties have been presented and integrated into an analytical framework. Starting from the basic definition of uncertainty as lack of information required to perform a task, three variations of this lack of information were discussed: incomplete information, inadequate understanding of existing information, and undifferentiated response alternatives. These different kinds of lack of information may concern the organization's and the environment's current or future state, cause-effect relationships, or available responses. Causes for uncertainties may lie within the organization, for instance, they could be related to technologies and materials used and the interdependencies between tasks, or they may be external to the organization, such as changing customer demands or emerging competitors. One main effect of uncertainty is lack of control, be it through lacking prerequisites for control in terms of transparency and predictability of the situation or through lack of clarity regarding possible means of influencing the situation. Finally, uncertainty may affect different actors and may have different consequences. In drawing up uncertainty landscapes for organizations, these different facets of uncertainty should be analyzed, leading potentially also to the exclusion of some facets which are not relevant for a particular organization or for a particular type of situation or decision analyzed. In the analysis, both objective characteristics of the organization and its environment and subjective enactments of the situation by different actors, including the sense-making processes involved, should be described. In an example from a railway company concerning a particular safety-relevant task the suggested framework for uncertainty analysis was illustrated.

3

Generic Principles for Managing Uncertainties

In Chapter 2, the aim was to show different facets of uncertainty in order to better understand the basic concept of uncertainty. In this chapter, approaches to managing uncertainties are focused upon. Quite obviously, there are very different demands on organizations both in terms of the uncertainties they are faced with and the requirements for turning uncertainties into business opportunities. As two extreme examples, take a company like Google and a nuclear power plant. While Google tries to achieve fast innovation in a very dynamic environment by giving the maximum possible freedom to bright young software developers, nuclear power plants are concerned with shielding their processes from uncertainty and maintaining highly professional routine operation for as long as possible. However, Google also needs certain structures and standards to ensure coordination within and between software development teams, whereas nuclear power plants face requirements for handling non-routine situations and for innovation in response to technological and regulatory developments. Analysis of internal and external uncertainties and decisions on how to manage them on a strategic and operational level are key to successful performance in both cases, even though the decisions taken will differ substantially.

As a starting point, minimizing uncertainties versus coping with uncertainties will be contrasted as two basic approaches (Grote 2004b). Subsequently, early literature on organization theory which aimed at developing contingency models for deciding between these two approaches in light of the types and amounts of uncertainty a particular organization is faced with, will be reviewed. More recently, research has been concerned with showing the need and also the possibilities for overcoming this dichotomy. Concepts such as loose coupling (Weick 1976; Orton and Weick 1990), enabling bureaucracy (Adler and Borys 1996), ambidexterity (Tushman and O'Reilly 1996) or semistructures (Brown and Eisenhardt 1997) have been suggested as possible ways to fulfil requirements of concurrent stability and flexibility that neither minimizing nor coping with uncertainty can accomplish alone. These different concepts will be defined and available evidence on how to achieve a balance between stability and flexibility discussed. Current knowledge on adequately managing uncertainty will be summarized by means of several generic principles. Moreover, a four-step decision process is proposed as a final generic principle that decision-makers in organizations should follow in order to better grasp requirements and possibilities of managing uncertainty and make design decisions accordingly. As part of this decision process, the usefulness of reducing, maintaining and increasing uncertainties is discussed both in light of internal and external contingencies that companies have to respond to, and in terms of basic belief systems developed, enacted and continuously recreated in companies in order to reduce fundamental uncertainties about appropriate management practices. At the end of the chapter, the railway example described in Chapter 2 is taken up again to illustrate the steps in the suggested decision process.

Before getting into the discussion about the different approaches to managing uncertainty, a general remark on the usefulness of prescriptions for organizational and system design is deemed necessary. With his seminal work on organizational design, Thompson (1967) has probably provided the most thorough account of a contingency approach to uncertainty, while at the same time trying to overcome the dichotomy of minimizing versus coping with uncertainty. In passing, he makes the interesting comment that the proponents of the minimizing uncertainties approach focus on designing efficient and effective organizations, in contrast to the proponents of the coping with uncertainties approach, who are interested in describing and understanding organizations. One may argue that the proposal to minimize uncertainties in organizations in itself contains a reduction of uncertainties assumed necessary to achieve design objectives. The proposal to cope with uncertainties on the other hand, tries to keep all uncertainties in the picture, but thereby renders prescriptions for design more difficult. Thompson's aim was to overcome the limitations of both approaches by building design propositions from a very complex model of the organization that fully acknowledges high levels of uncertainties. Though being much less ambitious than Thompson, who tried to formulate an all-encompassing model of organizational design, this book is faced with the same difficulty of balancing the simplification needed for design decisions and accounting for the complexity needed for a proper understanding of the relevant organizational phenomena.

The approach taken here for handling this difficulty is to start off with generic principles that can broadly guide decisions on practices used for managing uncertainties. In subsequent chapters, more concrete examples will be examined concerning design decisions in different organizational domains, specifically in the context of standardization and planning of work processes, automation, safety management, and employment relationships. One important advantage of framing design decisions within the suggested generic principles is that the compatibility of decisions across a variety of organizational and technological issues can be investigated, supporting a more integrated management approach. This will be discussed further in the concluding chapter of this book.

3.1 Minimizing Uncertainties versus Coping with Uncertainties

The major approaches to system design at the turn of the 20th century (Taylor 1911; Weber 1947) were built on the assumption that organizations are closed systems, thereby protected from external uncertainties. Concerning internal uncertainties, the assumption was that they should be minimized by minute planning and continuous monitoring of the execution of these plans, providing minimal degrees of freedom to the people in charge of carrying out the plans and taking any deviation from the plans as signs for the necessity of even more planning and monitoring (see Fig. 3.1). Accordingly, the basic control mode is that of feedforward control. The Fordist production lines are a prime example of the minimizing uncertainties approach. They were tailored to mass production of standard products, that is: Model T in black.

With the acknowledgement of the open system nature of organizations this approach continues to be followed and has even gained in fervor in order to keep systems under control: If the organization's environments become more and more complex – or rather the complexity is more and more acknowledged – then more effort has to be put into reducing the uncertainties connected with the complexity. As the minimizing uncertainties approach promises maximum control, it is still the favourite approach in many organizations.

Another approach which has been promoted by organization theorists and work scientists for several decades now is – instead of fighting uncertainties in an attempt to minimize the uncertainties themselves or at least their effect in the organization – to enable each and every member of an organization to cope with uncertainties locally and to allow for feedback control (see Fig. 3.1). From this perspective, planning is understood primarily as a resource for situated action (Suchman 1987), not as a blueprint for centrally determined and monitored action.

Local actors need to be given as many degrees of freedom as possible, achieving concerted action mainly through lateral, task-induced coordination. Disturbances – as much as they may be annoying momentarily – are also regarded as

Minimizing uncertainties	*Coping with uncertainties*
• Complex, central planning systems	• Planning as resource for situated action
• Reducing operative degrees of freedom through procedures and automation	• Maximizing operative degrees of freedom through complete tasks and lateral cooperation
• Disturbances as to be avoided symptoms of inefficient system design	• Disturbances as opportunity for use and development of competencies and for system change
▼	▼
Dependence / feedforward control	*Local autonomy / feedback control*

▲ *Balance through loose coupling* ▲

Motivation through task orientation
Higher order autonomy
Flexible changes between organizational modes
Culture as basis for coordination/integration

Figure 3.1. Basic principles of uncertainty management underlying organization design (from Grote 2004b)

opportunities for use and expansion of individual competencies and for organizational innovation and change. Cherns' (1976, 1987) principles of sociotechnical design provide a good summary of the core ideas of this approach, especially the principles of minimal critical specification for work processes and task allocation, of role breadth to ensure multifunctional expertise, and of controlling variances at their source. These principles argue for designing systems with much local decision latitude and overlapping competencies in order to cope with uncertainty locally.

3.1.1 Distribution of Autonomy and Control

Another way of looking at the two approaches to uncertainty is to describe them in terms of the distribution of autonomy and control in the organization. Differing from the frequent, almost synonymous, use of the two terms autonomy and control, they are used here to indicate two quite different types of influence (Grote 1997). Autonomy is defined as self-determination regarding goals and the rules to be followed in achieving these goals. Control is defined as the influence on a given situation allowing goals to be reached which have been determined either autonomously or by others.

This distinction is regarded as useful, especially for discussing the basic gaps between planning and implementation of plans in view of internal and external uncertainties (Grote 2001). When attempting to minimize uncertainty, autonomy stays completely with centrally located decision-makers who also try to maximize their control by prescribing in minute detail how the people implementing their plans have to use their influence in a given situation. Coping with uncertainty on the other hand, is characterized by maximum local control as well as sufficient local autonomy in order to choose or at least modify goals and rules for goal achievement in view of maximum effectiveness in the exertion of local control.

As will be discussed in more detail, especially in Chapter 5, the distinction between autonomy and control has been successfully employed, for instance, to study the interaction between production planning and scheduling and operations (Wäfler 2001, 2002), processes involved in demand forecasting (Windischer 2003), and relationships in supply chain management (Windischer and Grote 2003). In all of these cases, it was found that local control should be maximized, while autonomy should be distributed according to task interdependencies and goal agreement. When local actors can identify with goals, even though they have not participated in defining the goals, for instance, because there are shared interests across all parties involved, autonomy can stay with a central decision-maker such as a strategic network partner (Jarillo 1988). Also, when task interdependencies are sequential in nature, autonomy has to remain with a central planning body.

3.2 Contingencies in Managing Uncertainties

The distinction of two approaches to handling uncertainties, one focusing on reducing and the other on coping with uncertainties, is frequently found in organization theory research. One of the oldest and most well known examples is the work by Burns and Stalker (1961) on mechanistic versus organic organizations. While mechanistic organizations are characterized by high levels of specialization and differentiation, vertical unidirectional coordination, and hierarchical leadership, organic organizations are functionally integrated, with predominantly lateral and reciprocal coordination and network-based leadership.

Most of the work on different approaches to uncertainty management has not promoted one approach over the other, but rather various contingencies have been hypothesized and tested empirically. The most basic understanding of these contingencies is that minimizing uncertainties only works when the overall level of uncertainties an organization is confronted with is low (*e.g.,* Burns and Stalker 1961; Thompson 1967). With higher levels of uncertainties, any attempt to design them out of the system will fail and therefore the system has to be enabled to cope with uncertainties locally. This understanding lies at the heart of the socio-technical design principle of handling variances at their source (*e.g.,* Emery 1959; Cherns 1976).

Many studies have taken the contingency approach as a starting point to test specific hypotheses related to various aspects of organizational design (for a more comprehensive review see Wall *et al.* 2002). An early example is the study by Keller and colleagues (1974) in which the original distinction by Burns and Stalker (1961) between organic and mechanistic design was translated into three factors: impersonal hierarchy, group decision-making, and rules for decision-making. The authors related these factors to the amount of uncertainty determined by the number of major product changes and to economic success in continuous process plants. They found overall better performance of plants that had an organic design irrespective of degree of uncertainty, which was interpreted in light of the comparatively high level of task uncertainty in process plants in general.

In the study by Van de Ven and colleagues (1976) mentioned earlier (see Example 2.4. in Chapter 2), relationships between task uncertainty, task interdependence (independent, sequential, reciprocal, team) and coordination modes (impersonal coordination through policies and procedures, personal coordination between individuals on the same or different hierarchical levels, group coordination) were analyzed in a large state administration. In line with contingency assumptions, higher task uncertainty was related to less impersonal coordination and more personal coordination through group meetings and lateral communication. Also, personal forms of coordination increased and impersonal forms slightly decreased along the four types of task interdependence from independence through to team interdependence. Hierarchical personal coordination was stable across all conditions of uncertainty and interdependence. Overall the less costly impersonal mechanisms were used more and the more costly personal mechanisms, especially on a group level, were used least. Finally, task uncertainty seemed to be more important than task interdependence in determining coordination mode.

Again testing contingencies between task uncertainty and coordination modes, Argote (1982) studied work processes involved in hospital emergency admission. Her measure of task uncertainty was input uncertainty, that is, predictability of patient condition (see Example 2.1. in Chapter 2). She found that as predicted by contingency assumptions, programmed means of coordination (rules, scheduled meetings, authority arrangements) were related to effectiveness (promptness and quality of care) with lower levels of uncertainty, while nonprogrammed means of coordination (individual autonomy, general unit policies, mutual adjustment) were more effective with higher levels of uncertainty. Interestingly, though, no evidence was found that emergency units actually used those coordination modes more, that were more appropriate for their task conditions.

Some contingency-focused research has adopted a human resource management perspective, by focusing on different mechanisms for controlling human performance in relation to external and internal uncertainties (Eisenhardt 1985; Ouchi and Maguire 1975; Ouchi 1979). Specifically, behaviour control, that is direct surveillance of employees' behaviours while performing their job, and output control, that is measuring work outcomes such as production quantity and quality, were distinguished. Ouchi and Maguire (1975) related behaviour versus output control of employee performance to uncertainty as measured by incomplete understanding of means-ends relationships and task complexity using data from retail department stores. In line with contingency assumptions, they found that behaviour control was associated with better means-ends understanding and lower task complexity. The findings for output control were less clear and seemed to indicate mainly that output control is used by managers in response to being evaluated themselves based on output. Eisenhardt (1985) expanded these findings theoretically and empirically by introducing cost of behaviour and outcome measurement as additional constructs. Again she studied retail stores and found support for behaviour control being used with more programmable tasks. Also, behaviour control was more likely when behaviour measurement was readily available and outcome measurement was costly (*e.g.*, high administration costs for commission-based pay). Yet another mode of controlling human performance is input control, which happens via the definition of competences, skills, and attitudes that are prerequisites for being hired into a company. This control mode has been discussed by Ouchi (1979) as the mode most closely linked to the notion of building up resources for coping with uncertainties. These concepts will be taken up again in Chapter 8.

Sitkin and colleagues (1994) have raised awareness for uncertainty-related contingencies in total quality management. They argue that total quality management (TQM) often turned into what they term total quality control, stressing resource efficiency, performance reliability, error reduction, standardized statistical control, and constructive conformity. While this minimizing uncertainties approach to TQM may be appropriate when there are few uncertainties, it is highly inadequate in the face of high levels of uncertainties. Then, total quality learning is required based on skill exploration, slack resources, resilience, insight, non-standardized diagnostic information, and innovation. However, these assumptions were not put to an empirical test.

There are also a number of newer studies which have been carried out within a contingency framework, *e.g.*, concerning manufacturing flexibility (Anand and Ward 2004; Ketokivi 2006), project management (Faraj and Sambamurthy 2006; Levitt *et al.* 1999; Shenhar 2001), and organizational governance and control (Folta 1998; Hendry and Kiel 2004; Santoro and McGill 2005). Usually, more uncertainty has been found to be associated with more internal flexibility and less hierarchical forms of leadership and control, as would be predicted by the basic contingency model. Many recent studies are more specific in their approach and application area, taking the basic uncertainty contingencies more or less for granted. Some of these studies will be discussed in the following chapters, when management of uncertainties is dealt with in relation to standardization and routines, intra- and interorganizational planning, automation, safety management, and human resource management. Also, recent research has been concerned less with studying the dichotomy of minimizing versus coping with uncertainties, but rather focuses on exploring a more balanced approach in order to achieve both stability and flexibility in organizations. This research we turn to next.

3.3 Balancing Stability and Flexibility: The Concept of Loose Coupling

Thompson has probably provided the first definition of what later on has been termed loose coupling: "... we will conceive of complex organizations as open systems, hence indeterminate and faced with uncertainty, but at the same time as subject to criteria of rationality and hence needing determinateness and certainty" (Thompson 1967, p. 10). Thompson assumed that different parts of the organization are faced with different amounts of uncertainty and are also capable of handling uncertainties to varying degrees. Specifically, following Parsons' (1960) distinction of the institutional, managerial, and technical level of responsibility and control, he argued that on the technical level a maximum level of certainty needs to be achieved in order to secure effective performance, while at the institutional level, the organization is confronted with a multitude of environmental uncertainties over which it has only limited or no control. The managerial level has to mediate between these two extremes, passing on as few uncertainties to the technical level as possible, but also initiating changes at the technical level if response to uncertainties is unavoidable. This understanding is still visible in Weick's definition of loose coupling: "[...] the concept of loose coupling allows theorists to posit that any system, in any organizational location, can act on both a technical level, which is closed to outside forces (coupling produces stability), and an institutional level, which is open to outside forces (looseness produces flexibility). [...] This general image is described here as the dialectical interpretation of loose coupling." (Orton and Weick 1990, p. 205). However, Thompson assumed that different subsystems function either with a focus on stability or on flexibility and the balance is achieved on the level of the overall organization, whereas Weick argues that the balance between stability and flexibility must be achieved within each organizational subsystem. As will be seen later, this is still an open debate.

On a metaphorical level, loose coupling can be understood in terms of dynamic stability of a building, which is achieved by combining fixed structures and components with predefined degrees of freedoms. For earthquake protection, for instance, buildings are sometimes equipped with a large pendulum situated in an atrium or with large weights mounted on wheels on a track at the top of the building. When the building starts to shake, these weights will always move in the opposite direction, thereby counterbalancing the building's movement and increasing overall stability.

In Figure 3.1, four examples are given of loose coupling in terms of balancing minimizing versus coping with uncertainties in organizations:

1. The concept of *motivation through task orientation* (Emery 1959) assumes that tasks allowing for a high degree of autonomy, task completeness, and task feedback will further an individual's intrinsic motivation towards fulfilling the goals of the primary task of the organization, that is local control and autonomy will be used to achieve organizational goals.
2. The concept of *higher order autonomy* (Grote 1997; Klein 1991) has been suggested as providing autonomy in those situations where in technically tightly coupled systems (Perrow 1984) little operative autonomy is possible. Higher order autonomy allows the actors in the organization to decide on the restrictions of their own operative autonomy, *e.g.*, through participative design of rules and procedures (Bourrier 1998).
3. In Lawrence and Lorsch's study (1967) on differentiation and integration in organizations, it was found that different parts of an organization (R&D versus marketing versus production) can be organized differently in response to the different amounts of uncertainties with which they are confronted. More recently, in studies of high-reliability organizations especially (*e.g.*, LaPorte and Consolini 1991), it was observed that also within organizational units *switches between different organizational modes* are possible. Finally, even on an individual level, there is evidence that workers switch between modes of operation in response to situational requirements, for instance, between standardized production tasks and continuous improvement activities (Victor *et al.* 2000).
4. Weick (1987) has pointed out that *culture* serves as a form of coordination and integration that incorporates both decentralization of autonomy and centralization of values and norms as binding forces for local action. This has also been described as normative control, differentiating it from objective control through assessing performance against measurable standards and from self-management where – in Grote's (2001) terms – local actors have both high levels of control and autonomy (Leifer and Mills 1996).

3.3.1 Organizational Mechanisms for Balancing Stability and Flexibility

Much of the newer research in organization and management theory has explored the notion of loose coupling further by identifying and testing specific organizational mechanisms that provide concurrent support for stability and flexibility. Regarding manufacturing in particular, Gerwin (1993) analyzed at a conceptual level the requirements for managing environmental uncertainty. He distinguished four strategies that he labelled adaptation, reduction, redefinition, and banking. Adaptation involves responding to uncertainty by increasing flexibility, for instance, through a more diverse product mix, faster product modifications, or adaptable production capacities. Reduction describes the efforts a firm undertakes to reduce uncertainties and thereby also the need for flexibility, for instance, through long-term contracts with customers, extending product life cycles, or levelling demand. Most interesting are the other two strategies, redefinition and banking. Redefinition of uncertainty may actually involve an increase in uncertainty by creating higher customer expectations in terms of product innovation or delivery times. This strategy requires a previous increase in flexibility in order to handle the increased uncertainty. A major advantage of this strategy is the competitive advantage established, as other companies have to handle the increased uncertainty without being prepared for it. Finally, banking implies the building up of flexibility reserves that may then be used adaptively or in the course of redefining uncertainties proactively. All of these forms of flexibility help to establish stability on a higher level by securing the company's survival. The proactive forms of redefinition of uncertainty and banking flexibility allow a company to build up capacity for change without disruption of work processes and disorientation of the people involved. In a case study of 15 manufacturing units of a large corporation, Ketokivi (2006) found evidence for all four strategies and provided more detailed accounts of the contingencies involved in using them successfully. For instance, banking by building up extra resources for later use works well for specialized products with little price competition.

An empirical study of the actual tradeoffs between stability and flexibility was provided by Adler and colleagues (1999). They analyzed model changeovers in a Toyota plant, focusing on the tension created by having to provide for both efficient routine and innovation. Four mechanisms were assumed critical: metaroutines for the innovation process, such as standardized problem solving procedures and project reviews; partitioning of organizations into subunits specializing in routine versus non-routine tasks; switching between routine and non-routine tasks for individual workers; and enriching jobs to include both types of tasks in parallel. There was evidence for all four mechanisms in the plant, related to both the internal functioning of the plant and the organization of supplier relations. Training and trust were found to be the cornerstones for the working of the four mechanisms. Trust can counteract loss of organizational control implied by these mechanisms (Leifer and Mills 1996). Also, there were indications of a special form of bureaucracy that Adler and Borys (1996) named enabling bureaucracy, which – in contrast to coercive bureaucracy – entails more mutual adjustment in the coordination of work processes and involvement of workers in defining formalized procedures as well as in designing products and production processes.

In a study by Dean and colleagues (1992), tensions between decentralization and formalization were investigated with respect to effects of advanced manufacturing technology on organizational structure. Based on their finding that introduction of new technology was accompanied by both more formal rules on how to use the technology and more centralized decision-making during the actual work process, the authors concluded that formalization may be a mechanism for safely decentralizing. This can also be understood in terms of creating pooled interdependence in the decentralized units by working based on common standards, which reduces coordination requirements.

While the examples discussed so far describe general features of organizational structures and processes that help bridge the requirements of concurrent stability and flexibility, an example studied by Bigley and Roberts (2001), focuses on the real-time processes involved. They analyzed the functioning of the Incident Command System (ICS) in a fire department, which is a highly structured and proceduralized approach to handling crises such as, in this case, potentially devastating fires. They found three elements in the implementation of the ICS that appeared to be crucial for combining the seemingly opposing demands of bureaucratic structure and flexible crisis management:

1. Structuring mechanisms which allow the quick elaboration of structure when the ICS organization is built at the scene, adaptive role switching decided by the incident commander, decision authority migrating to the most capable people for particular decisions, and system resetting by disengaging the team from the task environment to redirect or reconfigure it when confronted with a new situation or failure of the strategy chosen initially.
2. Support for constrained improvisation by providing decision latitude especially to more experienced team members, and by allowing controlled adaptation or even violation of existing rules.
3. Cognition management methods helping to bridge the tension between structure and flexibility for each individual. These mainly involved spending much effort in developing and continuously communicating a shared understanding of the situation, and also carefully monitoring whether people still had sufficient cognitive resources to update their understanding of evolving situations. If needed, responsibility for keeping everybody aligned with the overall demands of the situation was explicitly reassigned to another person.

3.3.2 Demands on Stability and Flexibility with Very High Levels of Uncertainties

Taking IT product development as an example, Brown and Eisenhardt (1997) studied mechanisms that provide a balance of stability and flexibility in the face of particularly high demands for innovation in very volatile environments. By contrasting successful with unsuccessful firms, they found that successful product development was linked to clear responsibilities and priorities, and to extensive communication internally and with the external environment. This created a structure for efficient work processes and general orientation, but also sufficient leeway for improvised decision-making. Also, successful firms provided structure through carefully choreographed transitions between projects, including predictable time intervals based on regular release dates and prespecified prodecures for phasing team members in and out of projects. Finally, decisions on new products were based on probing the future rather than planning for the future. Small-scale experimental products, strategic alliances with customers and other technology firms, assigning responsibility for long-range visions to particular people, and regular meetings on possible future products provided a broad set of options and opportunities which allowed flexible and fast decisions on new product developments. The findings of Brown and Eisenhardt are an impressive illustration of what they termed semistructures. The core innovation process is left unrestrained, while supporting processes are more structured through assigning responsibilities and priorities and setting time frames.

Eisenhardt and Martin (2000) discussed the required balance between stability and flexibility in highly volatile environments in terms of dynamic capabilities. Dynamic capabilities are defined as the organizational and strategic routines by which firms achieve new resource configurations to match or even create market change (Eisenhardt and Martin 2000, p. 1107). This may involve, for example, integration of resources by combining varied skills and functional backgrounds to create a new product, or gaining new resources by hiring new people or building new alliances. Again, borrowing from a contingency perspective, the assumption is that high uncertainty requires more dynamic capabilities, which are "simple, experiential, unstable processes that rely on quickly created new knowledge and iterative execution to produce adaptive, but unpredictable outcomes" (Eisenhardt and Martin 2000, p. 1106). With lower levels of uncertainty, dynamic capabilities are closer to what has been termed organizational routines (Nelson and Winter 1982), which are repetitive patterns of activity that rely extensively on existing knowledge and slowly evolve over time. However, Eisenhardt and Martin also stress the need to provide some structure to help decision-makers make sense of situations and reduce anxiety in the face of high uncertainty. Simple rules may serve this purpose, such as Yahoo's two rules for forming alliances: no exclusive deals, basic service provided by the deal (*e.g.*, online greeting cards) must be free. Also, the fast creation of knowledge through prototyping helps, more so than reliance on implicit knowledge which takes a long time to build up and is by definition impossible to codify.

3.3.3 Balancing Stability and Flexibility in Internal and External Responses to Uncertainty

Gilbert (2005) studied newspapers' responses to the rise of digital media, distinguishing between changes in resource investment patterns and changes in organizational routines that use these resources. He found that major changes in financial and human resources allocated to print versus electronic media took place as well as customer dependencies being reconfigured. At the same time, internal routines became more rigid as authority was more centralized, experimentation was reduced, and there was an overreliance on existing processes and business models. Gilbert argued that this response pattern was mainly motivated by threat perception which led to adaptive change in responses to the external environment, and increased and maladaptive rigidity in internal responses. These findings led Gilbert to suggest that innovation might be better handled by structurally differentiating a new venture from its parent organization, which can then focus better on exploring opportunities and respond more flexibly both internally and externally.

This duality of flexibility and rigidity at different levels of an organization also exemplifies a basic tension in part-whole relationships described by organization theorists (Astley and Van de Ven 1983). Autonomy at one system level is always linked to constraints at another system level, and vice versa. Particular members of an organization can only act autonomously if people at higher levels of the organization are prepared to restrict their autonomy and delegate it to them. In Gilbert's example, the increased internal rigidity might help to foster external flexibility, or conversely, external flexibility might demand internal rigidity.

3.3.4 Balancing Stability and Flexibility from a Learning Perspective

In 1991, March wrote a very influential article on two modes of learning in organizations. He argued that there needs to be balance between exploration of new possibilities, concerned with search, variation, experimentation and risk taking, and exploitation of old certainties in terms of refinement, implementation, and efficiency. "Adaptive systems that engange in exploration to the exclusion of exploitation are likely to find that they suffer the costs of experimentation without gaining many of its benefits. They exhibit too many undeveloped new ideas and too little distinctive competence. Conversely, systems that engange in exploitation to the exclusion of exploration are likely to find themselves trapped in suboptimal stable equilibria" (March 1991, p. 71). In the competition for scarce resources in organizations, exploitation tends to win because benefits are more visible and short-term (Benner and Tushman 2003). As an example, March discusses the socialization of newcomers into organizations, pointing to the usual attempts in organizations to ensure fast learning of organizational routines in order to quickly reach efficient performance at the expense of the organization learning from the different viewpoints and prior experience of the new employee.

March's work has motivated much research into achieving a balance between exploitation and exploration and thereby also, between stability and flexibility (Gupta *et al.* 2006). Two key issues that have been discussed in this research are:

- need for concurrent exploitation and exploration, which has been termed ambidexterity (Tushman and O'Reilly 1996), versus phases in which one or the other form of learning are dominant, termed *punctuated equilibrium* (Burgelman 2002);
- pervasiveness of demands for both exploitation and exploration across all organizations versus a contingency perspective arguing for different demands on both modes of learning depending on contextual factors such as uncertainty.

Gupta and his colleagues (2006) argue that these two issues may be partially resolved by distinguishing between different levels of analysis and design, as was discussed earlier in terms of tensions in part-whole relationships (Astley and Van de Ven 1983). Ambidexterity may be achieved at the organizational level by having subunits that specialize in either exploitation or exploration and are integrated by senior management (Benner and Tushman 2003). Contrasting the notions of Lawrence and Lorsch's classic work on differentiation and integration, organizational models are suggested where exploitation and exploration subunits encompass all organizational functions, such as R&D, marketing, and production, but are organized differently in terms of their focus on minimizing or on coping with uncertainties (O'Reilly and Tushman 2004). Moreover, some organizations may specialize in routine production, while others specialize in the creation of new product ideas, which, when linked through proper alliances and supply chains, may result in balancing exploration and exploitation at an interorganizational level.

3.4 Deciding on a Management Strategy for Handling Uncertainties

The empirical examples and theoretical models presented above show the richness of current thinking about the problems of balancing stability and flexibility in organizations. To date, not one generally accepted framework either for understanding or for shaping loose coupling exists. However, there is agreement that most of the time decisions on managing uncertainties will have to integrate elements of minimizing and coping with uncertainties instead of following a clear preference for one or the other approach.

As generic principles, it can therefore be stated that

- minimizing versus coping with uncertainties are the two basic approaches to managing uncertainty;
- minimizing will only be successful if there are few uncertainties;
- for the most part loose coupling is desirable as a way to support concurrent stability and flexibility; and
- loose coupling requires a balance between minimizing uncertainty, which creates stability, and coping with uncertainty, which creates flexibility.

In the following, as a final generic principle, a decision process is suggested for evaluating and potentially reshaping the balance of stability and flexibility in a given company. At the end of this chapter, the railway example described in Chapter 2 is taken up again to illustrate the steps of the suggested generic process.

3.4.1 A Generic Process for Decisions on Managing Uncertainty

Before starting the suggested process, an analysis of uncertainties along the lines laid out in Chapter 2 has to be undertaken. The decision process described in the following aims to evaluate current handling of these uncertainties in an organization and to redesign the management of uncertainties based on this assessment. In order to do this, four steps are proposed, and these should be carried out by a team that includes the organization's experts of the domains to be covered (for instance, planners when suppy chain management is the focus of the decisions to be taken or process engineers when a new automation concept is to be developed), representatives of the people affected by the decisions (for instance, shopfloor personnel), and the actual decision-makers (for instance, the top management team or one representative from that team). Depending on how broad the chosen content domain is and how much documentation already exists on the work processes related to that domain, the four steps can be discussed within a few hours, or they might form the basis of a project running weeks or even months. If, for instance, the degree of standardization for the operative processes in one subunit of the organization is the focus, this can be evaluated within a day or two.

If the general philosophy of the organization is at stake in an attempt to turn it into an ambidextrous organization, the necessary assessments of current practice and changes needed may easily take months.

The suggested four steps for the decision process are the following:

1. Analyze costs and benefits of reducing uncertainty;
2. Analyze costs and benefits of maintaining or increasing uncertainty;
3. Explore belief systems in the organization related to managing uncertainties;
4. Discuss anticipated effects of the recommendations derived in steps 1 to 3 and repeat any step as necessary to come to a final decision.

The *first step* conforms to most classic treatments of uncertainty management and entails an analysis of costs and benefits of reducing uncertainty. The *second step* looks at the costs and benefits of the opposite strategy of maintaining or even increasing uncertainty. In order to find a proper balance between stability and flexibility, it is most likely that some uncertainties will be handled in one way and others in a different way. Therefore, the first two steps need to be carried out separately for all identified uncertainties. In these two steps, the relevance of the different analytic categories suggested in Chapter 2 will also become obvious. Identifying the source of uncertainty, for instance, is mainly important if one tries to reduce uncertainty because one has to work on that source then. The type of uncertainty (state, effect, response) is more important when one aims at coping with the uncertainty.

The basic assumption underlying the first two steps is that reducing uncertainty usually increases stability, while maintaining or increasing uncertainty supports flexibility. However, aiming to reduce uncertainty that would be better maintained can actually destabilize the system. For instance, if in production scheduling the sequence of orders is fixed with no decision latitude given to people on the shop floor to adapt it in response to local disturbances like machine breakdown, this may severely hamper the workflow in the affected unit and beyond. These kinds of considerations form the core of step 1 and 2 in order to define an appropriate balance between reducing, maintaining and increasing uncertainty that fosters loose coupling.

Steps 1 and 2 assume that rational, objective accounts of uncertainties and of ways of handling them are possible and are used as a basis for the assessments. In the *third step*, the perspective is switched to one of sensemaking and enactment. This perspective holds that, for decision-making, perceptions of uncertainty are more relevant than objective accounts, and that these perceptions and the actions derived from them are embedded in and shaped by decision-makers' belief systems. Either minimizing or coping with uncertainty may be the preferred way of managing uncertainty based on beliefs about control and trust. These beliefs will make certain responses to uncertainty more appealing than others, even though these responses may be quite ineffective due to the particular conditions. A manager of a non-profit organization may feel that people should be trusted to follow the humanitarian goals of the organization and therefore will provide them with much autonomy and decision latitude in handling uncertainties in their tasks. At the same time, these uncertainties might require managerial intervention because they challenge strategic positions of the organization (*e.g.,* substantial reduction in donations). If, based on the assumption of trust, no action is undertaken, this will most likely result in ineffective performance and frustration of all parties involved. This simple example may illustrate the importance of confronting presumed rational accounts of uncertainties and their management with individual and collective belief systems, as is done in step 3.

In the *fourth step*, an overall evaluation of the chosen and potentially readjusted strategy for handling uncertainties is carried out. The aim is to achieve a balance of stability and flexibility fitted to the particular needs of the organization. Such an evaluation is, of course, difficult to make. Instead of aiming for some absolute judgment, often it will be more helpful to define probing strategies for continuous re-evaluations of the achieved balance. In this step, the recursive nature of the whole process is also considered. The outcome of step 4 may be that no satisfactory way forward can be defined leading the discussion back to the previous steps. Also, during any of the other steps it may become apparent that previous steps need to be reworked.

3.4.2 Covering Strategic and Operational Processes

When going through the four steps, it is important to keep in mind that uncertainties may concern strategic and/or operational processes. Much of what has been written about the management of uncertainties has focused on decision-making at the strategic level. For example, Allaire and Firsirotu (1989) have distinguished three ways to handle uncertainties in strategic management: (1) the predict-and-prepare response; (2) the power response based on dominating or eliminating sources of uncertainties; (3) the structural response which entails building internal capacity for flexible response and adaptation.

Uncertainties on the operational level have been discussed much less in organizational literature. In Thompson's (1967) treatment of uncertainty, for instance, there is the general assumption that uncertainties on the operative, technical level should always be minimized. Mostly, handling operational uncertainty has been a topic in job design literature (for a summary see Wall *et al.* 2002). A main thread running through this literature is the debate on keeping uncertainties in the system in order to enrich jobs and develop competencies and motivation versus reducing uncertainties to maximize managerial control (*e.g.,* Slocum and Sims 1980). Putting it very simply, job designers assume that people on the operative level of the organization love uncertainties while managers hate them.

In the suggested process for making decisions on managing uncertainties, an attempt is made to cover uncertainties both on the strategic and the operational level. However, due to the diversity and complexity of uncertainties an organization may be faced with, this will only be possible through examples with no claim made for a comprehensive and systematic coverage.

3.4.3 The Four Steps in Detail

Step 1: Analyze costs and benefits of reducing uncertainty. As a starting point we make the assumption that the overall objective in individual and organizational decision-making is to gain and maintain control in order to achieve desired goals. Because for decision-makers in organizations uncertainty may in itself induce strong perceptions of threat beyond the actual threats of economic loss (Argote *et al.* 1989), the usual first reaction to uncertainties is to try to reduce them. Uncertainties stem from insufficient transparency and predictability of situations due to lacking or ambiguous information and from unclear options for taking action in these situations. Transparency and predictability can be created either by gathering more information or by eliminating the causes of opaqueness and unpredictability. This relates to the first two strategies mentioned by Allaire and Firsirotu (1989), that is, the predict-and-prepare response and the power response based on dominating or eliminating sources of uncertainties.

The core of the predict-and-prepare response is the collection of more information by a variety of means, such as forecasting techniques, computer simulations, or scenario building. Contingency plans are then developed on the basis of the predicted futures with the potential danger of only seeing those aspects of the evolving future that fit the predictions and neglecting the rest (*e.g.*, Brown and Eisenhardt 1997; Weick and Sutcliffe 2001).

The power response is rooted in a fundamentally different view of uncertainties: "don't predict the future, control it" (Allaire and Firsirotu 1989; Sarasvathy 2001). The basic approach is not to adapt to uncertainties, but to change the situation in order to eliminate uncertainties or to transfer them to others. Power has been described as the capability to shift uncertainties to less powerful others (Marris 1996; Pfeffer and Salancik 1978). Examples in supply chain management illustrate this approach: So-called strategic networks are created around a powerful hub firm that establishes influence over partners in the network similar to hierarchical relationships inside firms, for instance, by having suppliers produce exclusively for the hub firm. This reduces the uncertainty for the hub firm substantially with the remaining uncertainty often shifted to the supplier by making them responsible for supplying according to demand with any surplus due to reduced demand not being bought by the hub firm (Sydow and Windeler 1998; Jarillo 1988). Also, companies may buy themselves markets by buying other companies serving those markets successfully already (Allaire and Firsirotu 1989). Another power response is to rely on legal action such as antitrust actions or lobbying for specific regulations (*e.g.*, Hoffmann *et al.* 2008). As with the plan-and-prepare response, problems may arise because future developments are misjudged. However, the power response implies the additional advantage that costs of mistakes may be transferred to others, for instance, to suppliers who have to deal with the surplus production resulting from overly optimistic demand forecasts.

A completely different form of transfer of uncertainty is to buy insurance coverage. This approach will not be discussed further, though, as it is specific to particular kinds of risks such as liability, production loss, and environmental damage, which are not the main focus of this book.

As indicated by the contingency models discussed earlier, the more uncertainties there are, the more costly any reduction strategy becomes. Moreover, while the benefits of reducing uncertainty are quite obvious because control is increased, the costs are partially hidden. In particular, the loss of flexibility associated with uncertainty reduction is not always sufficiently taken into account. Other hidden costs may be caused by uncertainties not actually being reduced overall, but reduced only for certain actors by transferring them to other actors who may be very ill equipped for handling them. For instance, when uncertainties are shifted from the managerial level to the operative level by enriching jobs with responsibilities for internal coordination or customer contacts, along with the appropriate training and preparation, this may be very motivating as job design literature shows (*e.g.*, Slocum and Sims 1980). When uncertainties are shifted to actors without providing them with adequate means of control though, this is a main cause for stress (*e.g.*, Karasek 1979) and will most likely lead to inadequate handling of these uncertainties. A classic example of the latter is the introduction of new technology in order to streamline certain processes without adequately training the people that have to operate the technology.

Additionally, the interplay between strategies chosen to handle uncertainties in different areas of a firm's activities has to be considered. A classic example of the necessary trade-offs are decisions on R&D expenditures. High R&D investment implies uncertainties about the return on investment of the inherently uncertain innovation process, while avoiding this uncertainty may lead to increasing uncertainty due to losing competitive advantage. As Miller (1992, p. 326f.) phrases it, "not all uncertainty exposures should be reduced. Rather the firm should attempt to establish an uncertainty exposure profile that optimizes its returns for the risks assumed".

The general recommendation to be drawn is that reduction and control of uncertainties is frequently a desirable strategy, which, however, focuses perception on the expected. Thereby, threats as well as opportunities may be overlooked. Also, these strategies can be very costly due to resources spent on measurement and control of internal and external processes. Finally, they do require some degree of predictability and in the case of the power response, a very strong position vis à vis other actors. If these prerequisites cannot be met and/or there are good reasons to fear too narrow a focus on predictable and controllable futures, Allaire and Firsirotu's (1989) third approach is advisable, that is building internal capacity for flexible response and adaptation. This is the focus of the second step.

Step 2: Analyze costs and benefits of maintaining or increasing uncertainty. Miller (1992) was quoted above with his statement that firms need to establish an optimal uncertainty exposure profile instead of generally minimizing uncertainty exposure. Even more strongly, Jauch and Kraft (1986) have argued for increasing internal or external uncertainties as a viable strategic option. They point to research on prospector organizations whose role is to create uncertainty by providing information on market opportunities. Also, in the general problem solving literature, there is the recommendation to start with a phase of divergent thinking that opens up new possible avenues and thereby increases uncertainty before moving into a phase of convergent thinking aimed at deciding on a particular solution (*e.g.*, Brophy 1998). Similarly, in social psychology the establishment of a norm of criticism in decision-making groups has been advocated (*e.g.*, Janis 1982). Both of these examples hint at the more general issue that more information – usually discussed as a means to reduce uncertainty – may actually create new uncertainties if the information concerns events with unknown probabilities or allows for different interpretations (Becker 2004).

Gerwin's (1993) redefinition strategy for handling manufacturing flexibility mentioned earlier involves an increase in uncertainty by creating higher customer expectations. He argues that this strategy requires a prior increase in flexibility in order to handle the increased uncertainty. Thereby a competitive advantage is established, because other companies have to handle the increased uncertainty without being prepared for it. Some more examples of measures aimed at increasing uncertainty can also be taken from Brown and Eisenhardt's (1997) research on computer technology development. What they call probing the future, through small scale experiments and broad-scale discussions with customers, competitors and technology gurus, opens up arrays of options with highly uncertain viability. Finally, in their study of inter firm relationships, Beckman and colleagues (2004) point again to contingencies related to different ways of handling uncertainties. They found for the 300 largest US firms that companies tend to increase uncertainty through exploration by expanding their partner networks when firm-specific uncertainty is high (*e.g.*, due to a major reorganization), but reduce uncertainty by exploitation through strengthening existing networks when market uncertainty is high (*e.g.*, due to technological changes affecting a whole sector of industry). As firm-specific uncertainty is more controllable than market uncertainty, exploration seems to be chosen more with higher levels of control, while exploitation is a response to lack of control.

To date, there is little empirical research on the deliberate increasing of uncertainties due to the dominant view of uncertainties as inevitable, but largely unwanted. Even in the innovation literature, uncertainties tend to be acknowledged only to the extent that they are an unavoidable side-effect of discovery. Suggestions have even been made on how more routinized processes can be introduced to increase the predictability of innovation (*e.g.*, Brown and Eisenhardt 1995, 1997; Nelson and Winter 1982).

What is more frequently discussed is maintaining uncertainty and responding to this uncertainty by increased internal flexibility. "Unlike control and cooperation strategies which attempt to increase the predictability of important environmental contingencies, flexibility responses increase internal responsiveness while leaving the predictability of external factors unchanged" (Miller 1992, p. 324). A classic example of this strategy is diversification, be it with respect to products, markets or suppliers, which reduces resource dependence.

Peculiar to the dynamic capabilities that Eisenhardt and Martin (2000) discuss as appropriate for firms operating in high-velocity environments, is that no attempt is made to reduce uncertainty. Instead, the suggested "simple, experiential, unstable processes" maintain or even increase uncertainty as they mirror external uncertainties in order to provide requisite variety within the organization (Ashby 1956). Similarly, the notion of ambidextrous organizations implies that for those parts of the organization which are charged with producing highly innovative outcomes, uncertainties are maintained or increased (Benner and Tushman 2003).

Finally, the generic method of buffering by building slack and redundancies (*e.g.,* Staehle 1991; Thompson 1967) serves to acknowledge uncertainty and shield the organization against its effects. From a narrow efficiency standpoint, slack and redundancies are clearly to be avoided, as exemplified by the just-in-time production philosophy. However, many organizations have found that a very strict just-in-time production causes new uncertainties because flexibility is minimized (*e.g.,* Husseini *et al.* 2006).

Overall, there seems to be comparatively little to recommend maintaining or increasing uncertainties, unless radical innovation in highly volatile environments is the organization's main objective. The costs of acknowledging limited control, of increasing the variety in possible responses to external contingencies, and of building buffers are seen to easily outweigh the benefits of flexibility and responsiveness. However, this view is often skewed by overly optimistic beliefs about one's own influence and power. When assumptions about available options for reducing uncertainties turn out to be false, the cost-benefit calculation for maintaining or increasing uncertainties may start to look more favourable. Besides the rational approach to understanding and managing uncertainties, it is therefore quite important to be aware of the belief systems that individuals and groups develop with respect to uncertainties and their control and management. Different attitudes, norms and values associated with uncertainty may turn out to be more powerful determinants of decisions than the rational considerations presented so far. We now turn to these belief systems and their influence on decisions regarding the management of uncertainty.

Step 3: Explore belief systems in the organization. This book quite purposefully focuses on uncertainty, not on risk. While risk for most people has the negative connotation of something bad happening or something good not happening, the term uncertainty is more neutral, indicating a general lack of predictability or lack of understanding. However, some people may even consider unpredictability and ambiguity as risk in itself. One of the core dimensions for describing national cultures is in fact uncertainty avoidance (Hofstede 2001; Schwartz 1999). In some cultures, the lack of predictability about the future is perceived as a threat and people try to protect themselves against that threat by planning, by buying insurance or by staying as much as possible within the known. In other cultures, people do not mind the lack of predictability or even perceive it as something that adds excitement to their lives.

Similarly, one may distinguish companies as avoiding or embracing uncertainty. Reasons for these differences can be found in the surrounding national cultures – the American culture might be more conducive to furthering an entrepreneurial spirit than the Swiss culture for instance – but there are also tremendous within-country differences. One might try to explain those differences rationally by pointing to the necessity of either avoiding uncertainties, *e.g.,* in nuclear power production, or maintaining and increasing uncertainties, *e.g.,* in searching for new business opportunities, given the particular situation of a company. Step 1 and 2 emphasized this rational approach by focusing on costs and benefits of different strategies in view of different contingencies (*e.g.,* Thompson 1967). There is ample evidence though, that companies handle the seemingly same challenges very differently. This observation lies at the heart of the sensemaking and enactment school of thought (*e.g.,* Weick 1979, 1995; Salancik and Pfeffer 1978). Companies or rather, the individual actors in these companies, do not just react to their environment and the contingencies it presents, but they actively create their environment through processes of selective attention, enactment, and sensemaking. Even presumably rather objective characteristics such as task interdependence are influenced and (re-)created based on individual and group values (Wagemann and Gordon 2005). As Weick (1979) puts it very succinctly, meaningful environments are outputs of organizing, not inputs to it. "Managers proactively choose from their environment [...] what will be the relevant operating context for them. There may well be costs involved in ignoring certain contextual factors, but these are only costs to be weighed against alternative costs; they are not determinants" (Astley and Van de Ven 1983, p. 253).

Whether an uncertainty is seen as a risk or as an opportunity is at least partially in the eye of the beholder. Personal differences such as risk propensity (Sitkin and Pablo 1992; Nicholson 2005; Zuckerman and Kuhlman 2000) and differences in national cultures (Hofstede 2001; Schwartz 1999) influence this sensemaking process, but probably the strongest effects in organizations are related to the belief systems held by occupational groups and the belief systems developed in the companies themselves, that is, company culture. These latter two will therefore be expanded upon.

Concerning the influence of beliefs held by occupational groups, in Chapter 2, there was already some reference to the highly influential belief system embedded in the engineering-based Tayloristic approach to management. A core element of this belief system is the assumption of controllability of both technical and social systems based on objective, rational knowledge. Weitz and Shanhav (2000) postulated that engineers used their success in handling technical uncertainties to expand their professional domain to include the reduction and elimination of organizational uncertainties as well. How brittle this approach is, is shown by Feldman's (2004) analysis of two major NASA accidents; the explosions of the shuttles Challenger and Columbia. Feldman traces some of the faulty decision-making involved in these tragedies back to an over-confidence in quantitative data combined with neglect of non-quantifiable data. As an underlying cause, he sees the culture of objectivity at NASA, a culture he considers typical for an engineering organization. "They (the NASA engineers) were not able to quantitatively prove flight was unsafe, so in this culture it became easy for management to claim it was safe. [...] Under conditions of uncertainty, cultures dominated by the belief in [...] objectivity must be silent. This silence makes these cultures vulnerable to power and manipulation." (Feldman 2004, p. 708). He goes on to recommend changes in engineering training to support engineers in handling uncertainty also by non-quantitative approaches, in better understanding the social and historical roots of knowledge instead of taking it as an ahistoric and impersonal truth, and in speaking up against power and politics based on their professional expertise.

In another analysis of professional culture, Ferraro and colleagues (2005) investigate belief systems of economists and how they sustain certain theoretical assumptions and organizational practices. In particular, they discuss the core assumption of humans being motivated mainly by economic self-interest. They find that people choosing to study economics tend to adhere to this assumption of the "economic man" more than students in other fields and show how this assumption permeates economic theory despite evidence to the contrary.

Professional culture may be the driver for organizational culture if one profession is very dominant in an organization, as in the case of NASA. Often, organizational culture will develop from an amalgam of individual and group values and beliefs, however. Much has been written about organizational culture, about how influential it is and how difficult it is to grasp and change at the same time (*e.g.*, Schein 1985). This literature is not to be reviewed here, but just one basic statement is taken from it: Implicit beliefs, or what Schein calls basic assumptions, along with more explicitly expressed norms and values are powerful forces which shape decisions made in organizations. Very often, this is only a partially conscious process. Therefore, it is crucial for decision-makers to reflect – as much as this is possible due to their often implicit nature – on their beliefs and values before finalizing important strategic decisions. Uncertainty is such a fundamental fact of personal and organizational life, that beliefs and values concerning uncertainty are deeply ingrained in everyday activities. Therefore, to step back and try to understand why things are done the way they are is very important in order to understand past successes and failures better and to not consider certain strategies and tactics as to be taken-for-granted paths into the future.

An interesting approach to reflecting on value systems and their effects on decisions taken in organizations has been suggested by Quinn, (1988; Quinn and Rohrbaugh 1983). Starting from the question of what important indicators of organizational effectiveness are, these authors had organizational researchers compare different indicators according to their conceptual similarity. Multidimensional scaling techniques revealed three dimensions underlying the similarity judgments: (1) flexibility versus stability/control, (2) internal versus external focus, and (3) means versus ends. They then derived a competing values approach for organizational analysis from those dimensions, in which the means-ends dimension was further developed into two dimensions, namely employee support versus goal-attainment and innovation versus rules (see Figure 3.2). More recently, a survey instrument has been developed from the Quinn framework (van Muijen *et al.* 1999) that can be used to assess shared perceptions of these values in an organization. This can also be a useful exercise for a team when working through the steps suggested here because it helps to identify shared beliefs as well as differences in perspectives.

Given that Quinn and Rohrbaugh's (1983) starting point was definitions of organizational effectiveness, their approach implies that not only decision-making is influenced by value systems, but also the criteria used to evaluate the outcomes of those decisions. In the end, things turn out well or badly because based on deeply rooted beliefs about what is good and bad, we make decisions and act upon them in ways that create self-fulfilling prophecies: If my gut feeling tells me that a certain decision is wrong I will implement and enforce it meekly and thereby not allow its positive potential to come to bear; by the same token, decisions I strongly believe are right I will follow up with so much enthusiasm that I will mobilize others to join in to bring about the desired outcome; also, consciously or unconsciously I will use criteria for measuring the success or failure of that decision that fit my prediction.

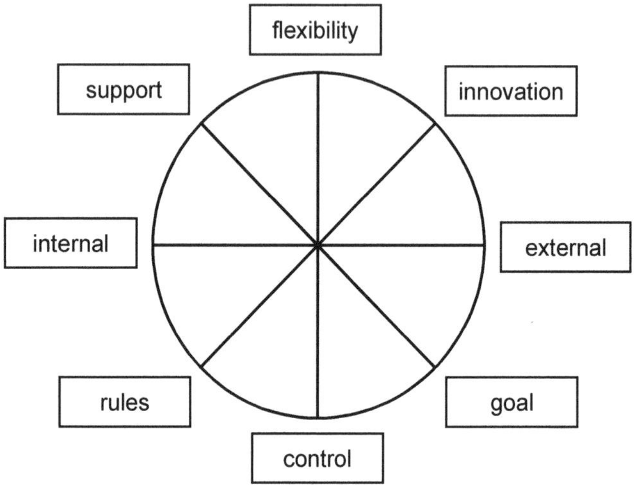

Figure 3.2. Quinn's competing values model (from van Muijen *et al.* 1999)

This is not only true for organizational practice, but also for theorizing about and researching that practice. "Social theories embody researchers' conceptions of causality" (Markus and Robey 1988, p. 595). Support for theories may be founded on self-fulfilling prophecies because when practices theoretically assumed beneficial are taken up by decision-makers they will do anything to prove the theory right (Ferraro et al. 2005, p. 8): "Theories can 'win' in the market place for ideas, independent of their empirical validity, to the extent that their assumptions and language become taken for granted and normatively valued, therefore creating conditions that make them come 'true'." Practically, these processes of self-fulfilling prophecies imply that it is quite difficult to objectively establish success or failure of a particular decision. Performance is as socially constructed as the criteria used to assess it. For instance, observing when the infamous "bottom line" is used as a criterion for or against a decision and when it is not, is an interesting case in point.

For many, this is probably too extreme a position to accept and one can surely find arguments for rationality and objectivity of decisions and their outcomes as well (Astley and Van de Ven 1983). However, to take on this position just for a moment to reflect on organizational action and its underlying reasons and desired and actual outcomes helps tremendously to gain a more thorough understanding of what it takes to manage better and to manage uncertainties better. It is hoped that this book provides convincing arguments for this claim and thereby supports sensemaking that prepares decision-makers for the unexpected (Weick and Sutcliffe 2001).

Step 4: Discuss anticipated effects. Steps 1 to 3 create an enormous wealth of information that does not necessarily reduce uncertainty about which strategy to follow, but more often than not increases uncertainty, at least in the short-term. The team having worked through these steps, has produced lists of uncertainties the organization is facing along with recommendations for their handling in terms of costs and benefits of reducing, acknowledging or even deliberately increasing them. There has been a thorough questioning of these recommendations regarding the viability of the underlying beliefs and values and potentially also, modifications in order to correct for wishful thinking or its opposite. Explicitly or implicitly, there has been reference to presumed outcomes of decisions all along. In step 4, an explicit overall assessment should be attempted based on the fundamental question as to whether the chosen measures for handling uncertainties achieve the ultimate goal of self-control through the appropriate balance of stability and flexibility.

When making this assessment, one should also consider Green and Welsh's (1988) argument that part of the strategy for managing uncertainties may not be to increase control over uncertainties, but to reduce the need for control. Taking uncertainties stemming from resource dependence as an example, this may even imply changing the organization's goals in order to reduce the amount of resources needed from a relationship, *e.g.,* lowering quality standards to be met by a supplier. Clearly, such a response does not increase control over the relationship. Instead, goals are adjusted to reduce the need for control, potentially as a consequence of admitting to insufficient power and influence in this relationship. A possible danger of this approach is that it may shift power gradients in the relationship even further. On the other hand, giving up control may have positive effects by indicating trust in a relationship. The more this trust is expressed from a general position of power, however, the stronger the positive signal will be.

Furthermore, it has to be kept in mind that the notion of control itself may be embedded in different belief systems. Therefore, it is very important to develop a shared understanding of control in an organization. Is it about knowing and influencing everything, is it about trusting other people in using their means of control, is it about reducing the need for control by loosening dependencies? The competing values approach by Quinn and Rohrbaugh (1983) presented earlier is a very useful instrument for exploring different understandings of control. Because desires for and use of control and – closely related – power are very important drivers in organizations, this issue will be taken up again in the next section of this chapter.

Finally, in step 4 the recursive nature of the whole process is considered. The outcome of the assessments made may be that no satisfactory way forward can be defined leading the discussion back to the previous steps. Of course, during any one of the other steps, it may also become apparent that previous steps need to be reworked.

3.5 Power and Control in Relation to Managing Uncertainties

So far, we have looked at the management of uncertainty within a general framework of control, where control is understood to be the ability of an individual or a system to influence situations for achieving certain goals. Some authors have added spice to this rather neutral description by arguing that the ability to cope with uncertainties defines power in organizations. From this viewpoint, coping with uncertainties may imply reducing as well as retaining or even increasing uncertainties.

In their "strategic contingencies' theory of intraorganizational power" Hickson and colleagues (1971) postulated and subsequently found supporting empirical evidence (Hinings *et al.* 1974) that subunits in organizations are more powerful the more they are able to cope with uncertainties. Power is reduced when the subunit's activities can be substituted by other subunits, that is, when their ability is not unique, or when the subunit's activities are not central to the organization's primary task, that is, their control concerns uncritical resources. Routinization reduces power by either reducing uncertainties as such or by defining ways for handling the uncertainties which increases substitutability. To this day, a frequently cited example to support these claims is a study by Crozier (1964), where the maintenance department was found to be the most powerful subunit in a cigarette factory because they controlled the one main uncertainty left, that is, the repair of machinery. This study is also a good example of Hickson and colleagues' point about routinization because the maintenance engineers secured their power by resisting routinization: Documentation of machinery was made to disappear and new engineers were only trained verbally.

Pfeffer and Salancik (1978) have pointed out two important restrictions to the general validity of the contingency-power relationship. One of these restrictions concerns the fact that power does not only develop from controlling contingencies, but that power can also be used to shape these contingencies. Again, the maintenance engineers in Crozier's study are an example of that – they actively resist routinization and thereby the threat of being substituted by other subunits or by external specialists. The other restriction stems from institutionalization of power through structures and policies that favour one subunit's influence over others. Institutionalization may also be a sign of the power of subunits that can bring about such policies.

Other authors have linked power to the control of uncertainties with special emphasis on the power to transfer uncertainties to others less powerful (Marris 1996; Pfeffer and Salancik 1978): While Pfeffer and Salancik postulate a need for more centralization of power in order to coordinate the control of uncertainties better and thereby avoid undue transfer of uncertainty, Marris argues for collaborative management of uncertainty supported by agreements and contracts that further the norm of reciprocity (Gouldner 1960). The notion of collaborative management of uncertainty will be elaborated upon specifically in Chapter 5 with respect to planning within and across organizations.

In their impressively comprehensive account of power in organizations, Clegg and colleagues (2006) again mention the ability to deal with uncertainties as an important source of power, in as much as unmanaged uncertainties could impede control. Their concern is more generally the analysis and critical assessment of power relationships in organizations though, thereby leaving the stance of older organization theory behind that organizations, by definition, are legitimized authority systems. With respect to the purpose of this book, this concern translates into analyzing uncertainties in the context of power relationships and the individual actors' capabilities for controlling uncertainties. Existing use of these capabilities by different actors should be evaluated in view of an optimal mix between reducing, maintaining and increasing uncertainties and the avoidance of undue transfers of uncertainties.

Such an evaluation should be undertaken for whatever work processes are at the centre of attention in a particular strategic or operational design decision, but it is also of great importance for decisions on the overall governance system of an organization. An interesting example in this respect is the research by O'Mahony and Ferraro (2007) on the development of a governance system in an open source community. When autocratic leadership emerged early on, it met with great resistance, leading to a long discussion process within the community on the bases for formal authority. With increasing needs for deliberate design of internal structure and for streamlining activities towards achieving agreed goals, formal authority built on management capabilities and kept in check by democratic rules was finally accepted.

In the remainder of this book, this very broad perspective will not be pursued further, as the focus is on the design of specific functions and processes within organizations. But it is worth pointing out that the principles discussed and the design process suggested might also be applied to decisions on organizational governance at a more general level. Judging from the current turbulences in the financial systems, the organizations involved are in great need of re-evaluating power relationships and the handling of uncertainties within these relationships, both on the level of single functions and processes and on the level of chosen forms of governance.

3.6 The Railway Example Revisited

In Chapter 2 uncertainties involved in shunting a train onto an occupied track were analyzed. We saw that this task contains a variety of uncertainties for the two actors concerned, that is, the train driver and the signaller. The motivation for conducting the analysis came from several incidents in which it had become clear that these uncertainties were not optimally handled. In the following, the suggested decision process will be used to explore options for better managing these uncertainties. Also, the concrete decisions taken at the railway company will be presented.

Step 1: Reducing uncertainties. There are two possibilities for reducing uncertainties by technical means: change the dwarf signals to have one unambiguous sign configuration per state or action required, and show the moving train on the signaller's track display instead of just showing occupation of track sections. Improved signals would eliminate most of the uncertainty for the train driver but are far too costly. An indirect measure for reducing ambiguity of the "go with caution" signal would be to reduce response options by enforcing a generally lower speed in combination with increasing awareness for the possibility of meeting an obstacle. Having a better display would provide the signaller with more information on the train's speed, but would require the signaller to pay close attention to the display and communicate with the train driver immediately if he or she drives the train too fast. Again, this solution is costly and also, paying more attention to single trains is difficult given the signaller's high work load due to very dense traffic.

The most obvious organizational option for reducing uncertainties is to enforce the recently introduced change of assignment of shunting supervisor to the train driver. The role of shunting supervisor entails determining the track to be used, informing all persons involved of the planned train movement, and requesting a go signal for the chosen track. If train drivers were to take on this role properly, uncertainties for them would be reduced because there would be full information on the status of the track used. Depending on the overall situation in which decisions on tracks have to be made, this role can encompass new uncertainties, however, as train drivers may not have full information on train movements in the station. For the signallers, uncertainties would remain the same or also be increased, because they would have less leeway in assigning tracks, which they particularly need during times of dense traffic.

Another organizational option for reducing uncertainties would be to require the signaller to inform the train driver via personal communication about the assigned track and the possibility of the track still being occupied when the shunted train gets there. During times of dense traffic, signallers may not have the extra resources needed to provide this information, though.

Finally, as a very fundamental measure, traffic density could be reduced, but this is considered unfeasible due to increasing mobility demands.

Step 2: Maintaining/increasing uncertainties. Upon first analysis, there is nothing to be gained from increasing uncertainties in this particular case. One may however, stretch imagination by investigating analogies to the free flight scenarios discussed in air traffic management (e.g., Sherali *et al.* 2000), which entail the pilot – instead of air traffic control – managing route selection and aircraft separation with the hope of more flexible and more effective use of air space based on pilots' local knowledge. Applying this scenario to railways, this would allow options to be explored for train drivers to take on more tasks in selecting tracks and speed, which increases uncertainty due to more decision-making requirements and higher task complexity for the train driver. The step taken by the railway company of reassigning the task of shunting supervisor from signaller to train driver is, in fact, a move in that direction. This step was taken because, due to increasing centralization of traffic control, only the train drivers still have local knowledge. To move even further in that direction would require major and thereby also expensive, technological changes, for instance, a screen in the engine cockpit showing the current state of track occupation in a station.

Maintaining the uncertainties that directly affect the train driver in the current situation, that is, incomplete information regarding track occupation and required speed – and thereby sticking to the old role assignment of shunting supervisor to the signaller – carries the advantage of maximum decision latitude for the signaller for handling uncertainties stemming from dense traffic. Also, maintaining the uncertainties implied in the unclear role assignment may allow adaptive coordination in line with situation-specific demands of particular shunting operations.

Step 3: Belief systems. In line with the strategic contingencies theory described earlier (Hickson *et al.* 1971), signallers are considered to be the most powerful actors as they need to cope with the most, and also with the most critical uncertainties. Shifting the role of shunting supervisor from signaller to train driver changes the power distribution for this particular task. The new role assignment may not have been implemented in part because it works against the current belief system. This interpretation is supported by the fact that train drivers were poorly instructed regarding their new role.

Step 4: Anticipated effects. All of the technical solutions are far too costly for the expected benefits. Requiring the signaller always to inform the train driver personally about decisions taken and their consequences for the train driver is not deemed realistic due to signallers' high task load. Furthering an awareness that obstacles always have to be expected and therefore shunting speed should generally be lower, does not reduce the uncertainty per se, but reduces negative consequences, as accidents are less likely. However, the lower speed may cause new problems in cases of very dense traffic and related time pressures. Finally, there are different advantages and disadvantages of the old and the new role assignment of shunting supervisor, as well as leaving the role assignment somewhat open as was de facto the case. Most likely, some mechanism for explicitly assigning the role to either train driver or signaller in specific situations would provide decision latitude for situated handling of shunting operations without creating unproductive ambiguities in the coordination between signaller and train driver.

Actual measures taken. Signallers are no longer required to inform train drivers when the track to be moved to is occupied. Instead, train drivers are instructed always to expect an obstacle after a "go with caution" signal and therefore generally reduce shunting speed. This new rule does not reduce the train driver's state and response uncertainty regarding the status of the track to be moved to and the correct speed for getting there, but it assigns the responsibility for handling this uncertainty more clearly to him or her. Additionally, some stations were determined where the general maximum shunting speed is 30 km/h instead of 40 km/h due to complex track and signal arrangements that reduce visibility. This reduces the pressure on the train drivers to drive faster in response to demands made by signallers and to the general push for punctuality. Finally, the role assignment of shunting supervisor to the train driver is enforced with one adaptation, which concerns requesting a track by the shunting supervisor. When both the track where the shunted train currently is and the track to which it is supposed to be moved are known to the signaller, the signaller can set the signal without request by the train driver.

3.7 Chapter Summary

In this chapter, principal ways of managing uncertainties have been discussed, especially the opposing strategies of minimizung versus coping with uncertainties and the notion of loose coupling as a way of combining advantages of the two strategies in order to achieve a balance between stability and flexibility. Newer research on concepts such as enabling bureaucracy, semistructures, and ambidextrous organizations was assessed with respect to its theoretical and practical value for achieving that balance. Current knowledge on adequately managing uncertainty was summarized by means of four generic principles: minimizing versus coping with uncertainties are the two basic approaches to managing uncertainty; minimizing will only be successful if there are few uncertainties; mostly loose coupling as a way to support concurrent stability and flexibility is desirable; loose coupling requires a balance between minimizing uncertainty, which creates stability, and coping with uncertainty, which creates flexibility. Subsequently, as a final generic principle, a decision process was suggested for evaluating and potentially reshaping the balance of stability and flexibility in a given company. This process consists of four steps: (1) analyzing costs and benefits of reducing uncertainties, (2) analyzing costs and benefits of maintaining or increasing uncertainties, (3) exploring belief systems in the organization, (4) discussing anticipated effects and repeating any of the first three steps if necessary. Because, in practice, uncertainty reduction is often the favoured strategy, particular attention was paid to costs of this strategy and to benefits of maintaining or even increasing uncertainty. Also, the importance of belief systems was pointed out as a means of influencing decisions on uncertainty management in a possibly non-rational and also not fully conscious manner. Finally, power in organizations was briefly discussed because the control over uncertainties can be an important source of power. Therefore, in the suggested decision process, the assessment of different ways of handling uncertainties should always take existing power relationships and possible changes in these relationships due to changes in the management of uncertainty into account. At the end of the chapter, the railway example first presented in Chapter 2 was analyzed further along the four steps suggested and the concrete measures taken by the decision-makers in the railway company were described.

4

Coordination in Organizations: Creating Flexible Routines

"Without the specification of the lines and areas of responsibility and the methods of control, it is difficult, if not impossible, for any social structure to direct its sub-organizations and members towards the achievement of the organization goals. These specifications may be formalised or they may be carried on in the cognitive structures and goals shared by the persons concerned. The attempt to formalise all possible uses of delegated powers is a fruitful source for 'red tape'. 'Over-structuring' in this respect can create a situation where the organization goals are lost sight of in a maze of prescription [...] and newly emerging problems are the responsibility of no-one and the playthings of whoever cares to go beyond his rigidly prescribed power." (Emery 1959, S. 38)

This quote by Emery is, even fifty years on, one of the most succint descriptions of the difficulties involved when using formalized rules as the basis for coordinated action. The quote also hints at another form of coordination, that is coordination through shared goals, which might be extended also to shared norms and values as embedded in an organization's culture. Determining adequate forms of coordination, along with deciding on the division of labour in an organization, is the most fundamental issue in organizational design. Most simply, coordination can be understood as managing dependencies. More elaborate definitions describe coordination as the activities needed to integrate or link together different parts of an organization to accomplish a collective set of tasks (Van de Ven *et al.* 1976, p. 322). Usually, five broad categories of coordination mechanisms are distinguished (*e.g.,* Thompson 1967; Van de Ven *et al.* 1976):

- technologically defined processes,
- central programmes and plans, *e.g.,* standard operating procedures,
- personal leadership,
- mutual adjustment via reciprocal team interaction, and
- cultural norms.

The ordering of these coordination mechanisms follows the rigidity of resulting action and thereby lack of flexible responsiveness to changing circumstances mentioned in Emery's quote. The most rigid forms, that is programmed impersonal modes of coordination, should be used with low levels of uncertainty, while more flexible nonprogrammed, personal modes should be used with higher uncertainty (*e.g.*, Argote 1982). Furthermore, the more personal forms of coordination are considered to be more costly and therefore should be used less overall (Van de Ven *et al.* 1976). So, we are back to the fundamental question of striking the right balance between stability on the one hand and flexibility and openness to change on the other in light of the uncertainties the organization is faced with. With respect to coordination, this question can be rephrased in terms of requirements for adaptive coordination, that is, switching between different coordination mechanisms in line with situational demands (*e.g.*, Entin and Serfaty 1999; Faraj and Xiao 2006; Grote *et al.* 2004, 2009; Manser *et al.* 2008, Zala-Mezö *et al.* 2009). In this chapter, we will approach this question by focusing on rules and routines as a means of coordinating work processes.

There is a long history of studying rules in organizations originating from Weber's work on rules as a core characteristic of bureaucracies (Weber 1947). Based on the notion that bureaucracies are very rigid forms of organization, rules and more generally, standardization, have acquired the negative connotation of stifling initiative and innovation (Stinchcombe 2001). Also, rules have been discussed as the key principles underlying hierarchical control and power in organizations (Clegg 1981). However, more recently, research has accumulated that argues against this one-sided view on rules by showing how rules may support flexible action. Often, this research focuses on organizational routines rather than on rules based on the understanding of routines as more generic forms of standardized behaviour. The concept of flexible routines (Howard-Grenville, 2005) takes into account that routines must be enacted by people to come to life and that this enactment process inevitably allows for variation and change in the routine (Feldman and Pentland 2003). The purpose of this chapter is to establish links between organizational routines and management of uncertainties, which should prove useful in resolving the dilemma of concurrent stability and flexibility. Using the literature on organizational routines, concepts for the design of rules and rules management as sources for loose coupling will be described. Subsequently, a case study on rules management in a railway organization will be presented and discussed in terms of propositions concerning contingencies for the effects of rules. Finally, implications for building and handling organizational routines within a framework of systematic rules management are discussed, integrating also the decision process for managing uncertainties introduced in Chapter 3.

4.1 Organizational Routines

Organizational routines have been defined as "repetitive, recognizable patterns of interdependent actions, carried out by multiple actors" (Feldman and Pentland 2003, p. 95). This definition points to routines as a basis for coordinated action, without specifying in which form these routines exist, for instance, as written-down rules, technologically determined courses of action, or experience-based tacit understandings of the right course of action. The basic assumption is that routines develop in organizations because they reduce complexity and uncertainty and increase stability, managerial control, and legitimacy. Routines may even support innovation, especially if innovation involves new combinations of known products and processes. Also, the handling of a non-routine condition in one part of an organization might be part of the routine operation of other personnel, such as maintenance personnel fixing machine problems (Nelson and Winter 1982).

More specifically, Nelson and Winter (1982) have described three functions of routines:

1. Routines embody *organizational memory* of the knowledge needed for successful task performance. The knowledge is kept alive by exercising the routine. Routines in this understanding support coordination based on shared goals and expectations without the need for explicit coordination between the individual actors. Reliability and consistency of action are achieved, because routines involve coordination without surprises.
2. Routines operate as *targets* for behaviour, when the routine's order is imposed on actors, thereby keeping behaviour under control. Routines may also be a target for replication and imitation when they have been considered conducive to success in the past, be it in one's own organization or in another, possibly competing, firm.
3. Routines help to establish a *truce* between conflicting interests of different participants in the organization because routine operation may involve routinely occurring laxity or even rule-breaking, along with routinized control mechanisms, keeping the overall routine functional and effective. The truce itself is not a fixed state, but needs to be re-established continuously, for which routinized patterns of action are used, however.

The case study mentioned earlier on incident command systems in a fire department (Bigley and Roberts 2001) contains a good illustration of routines as truce. The authors describe the degree of improvisation allowed which may even involve violation of standard operating procedures if the situation demands this and the individual in question possesses sufficient experience. Also, routines get adapted when individuals from several companies have to coordinate. However, "improvisation is considered legitimate and supported only to the extent that it fits with extant organizational goals" (Bigley and Roberts 2001, p. 1289). Based on an analysis of rule violations by shunters in a railway organization, Lawton (1998) points to the importance of supervisors for maintaining informal group norms which keep rule violations within the bounds of overall goal attainment.

4.1.1 Flexibility of Routines

In the literature, routines have come to be regarded mainly as a static product of learning, which – once established – impede further learning and thereby reduce organizational flexibility. The delicate balance between allowing variations in routines and enforcing routines as part of the truce described above may also reduce flexibility, as "adaptations [...] may be foreclosed because they involve a perceived threat to internal political equilibrium" (Nelson and Winter 1982, p. 111). Moreover, the control systems used to protect against unwanted change in routines might impede adaptation even when adaptation is clearly necessary.

Feldman and Pentland (2003) challenge the prevailing view of routines hindering organizational flexibility by arguing that routines always contain the duality of principle and practice. On the one hand, the principle of a routine as determined by a written procedure, a taken-for-granted norm, or some shared procedural knowledge has to be put into practice and in this process adapted to the necessities of a concrete situation. Rules are resources for action, but they do not fully determine action (Feldman and Pentland 2003, p. 101). Similarly, Reynaud (2005) argues that rules are inherently incomplete due to their general and abstract nature. To fill this "void at the heart of rules" (Bourdieu 2005; Reynaud 2005), specific and contextualized routines need to be enacted. In this enactment process, the routine in principle helps by guiding, accounting, and referring. Guiding is accomplished by the routine serving as a normative goal for action. By providing explanations for what we do, routines also support accounting for actions. Finally, routines can provide simple labels for complex action patterns, which can be used as a commonly understood reference to these sets of actions.

On the other hand, the routine in practice is essential for the establishment and maintenance of the routine in principle, as routines only develop through repeated action. At the same time, the routine in practice can also modify the routine in principle when new ways of acting are found to be appropriate under specific circumstances. Whether these modifications get incorporated in the routine in principle depends, for instance, on the power of the respective actors to turn exceptions into rules. Enacting routines may therefore also be the source for change and flexibility. However, the exact preconditions under which the enactment of routines leads to stability or change are not known, as Feldman and Pentland (2003) state.

This way of conceptualizing organizational routines incorporates static as well as dynamic elements, or in other words, an ostensive aspect (the more or less static routine in principle) and a performative aspect (the routine in practice, which inevitably allows for improvisation and flexibility). This perspective of a duality of principle and practice may be useful for solving the dilemma of the concurrent necessity for both stability and flexibility in the everyday operations of organizations.

It is important to point out that this duality refers to the "inherent capability of every organizational routine to generate change, merely by its ongoing performance" (Feldman and Pentland 2003, p. 94). This is to be distinguished from the discussion about "dynamic capabilities" which are thought of as meta-routines serving the organization's strategic adaptation to changing environmental demands in the sense of "a learned and stable pattern of collective activity through which the organization systematically generates and modifies its operating routines in pursuit of improved effectiveness" (Zollo and Winter 2002, p. 340). This distinction needs to be understood in terms of differentiating between strategic decision-making on organizational change and the operative balance between stability and flexibility in everyday work processes. The focus in this chapter will be on the latter.

4.1.2 Conditions for Flexible Use of Routines

Several studies have addressed factors that influence flexible use of routines and routine changes. Analyzing the use of project management routines at a chip manufacturer, Howard-Grenville (2005) suggested that actor's time orientation (to past, present or future) and the degree of embeddedness of a given routine in other organizational structures, influence the flexibility with which a routine is applied and how likely the routine will be changed as a consequence of this flexibility. She assumed that present and future orientation and weak embeddedness further flexible use and change, while strong embeddedness hinders change – though not flexible use – even with present and future orientation. Edmondson and colleagues (2001) described leadership and team processes associated with the change of routines in surgery teams required by the introduction of minimally invasive cardiac surgery. This technology increased task interdependence and the surgeon's role shifted from that of an order giver to a team member in the more interdependent process. They reported that involvement, which furthered the teams' motivation, preparatory practice sessions and early trials, which both helped to create psychological safety while encouraging new behaviours, and reflection, which promoted shared meaning and improvement initiatives, were crucial for the successful establishment of new teamwork routines. Gilbert (2005) looked at company-wide changes in response to new environmental demands, using newspapers' responses to the rise of digital media as an example. He found that pressures for change resulted in new patterns of external resource use, but increased internal routine rigidity, that is, more centralized authority, less experimentation and focus on existing internal resources.

Routines may also be more or less beneficial depending on the task at hand. In a study by Gilson and colleagues (2005) the effects of standardized work processes and support for creativity were analyzed regarding two outcomes of technical service teams, that is, technical performance and customer satisfaction. Standardization was positively related to customer satisfaction, while creativity was positively related to technical performance. These findings indicate that routines do not necessarily have a general effect, but might be appropriate only for achieving some outcomes. In this particular case, it could be argued that good technical performance requires flexibility in handling uncertainties, while customer satisfaction depends on establishing a sense of stability and control.

Becker and Knudsen (2005) postulated that routines can be especially helpful when decisions must be made under pervasive uncertainty, where neither all the different outcomes nor their probabilities are known and information is highly ambiguous. Using routines under those circumstances, as they point out, does not reduce the actual uncertainty, however, but rather the amount of uncertainty decision-makers experience because orientation and a sense of control are instigated. Similarly, Zollo and Winter (2002) suggest that knowledge codification is particularly important for infrequent and heterogeneous tasks where decision-makers may experience the most unease.

4.1.3 Routines in High-Risk Systems

The effects of routines in the form of rigid standards have been discussed particularly in view of the demands in high-risk systems. Standard operating procedures have been developed in these organizations with ever increasing detail in order to streamline human action and to reduce its influence as a risk factor. Procedures are often a direct consequence of incidents and accidents, the analysis of which provides knowledge of unforeseen wrongful courses of action against which new rules are developed as a defense. While generally there is an understanding that rules are useful guides for safe behaviour, there is also an increasing concern that too many rules incrementally developed will not make up a good system to help human actors do the right thing especially in states of abnormal operation where they would need strong, but also flexible guidance (*e.g.* Gilbert *et al.* 2007; Dekker 2003; Woods and Shattuck 2000).

These concerns go back to basic observations on how rules specifying the exact operations to execute can have a detrimental effect on action because they do not allow the performing person to develop a plan of their own, but instead further the atomization of actions and the focus on micro-difficulties (Vermesch 1985). Also, people may violate rules intentionally because they do not see the particular rule fitting the purpose of getting a job done satisfactorily (Mearns *et al.* 2001); this has also been termed pro-social rule breaking (Morrison 2006). Then, there is the opposite problem that reliance on common standards may turn into an over-reliance, impeding switches to more explicit coordination and to higher levels of common action regulation, that is, switches from skill-based to rule-based or to knowledge-based behaviour. This problem can be exacerbated by the fact that standardization is a strong force towards shared understanding of a situation and its demands in a team, because it creates a common framework for team behaviour, reducing the need for explicit coordination. The expectation of shared goals, plans, perspectives, and knowledge bases created by reference to the same set of standard operating procedures, as helpful as it is under most conditions, does involve the risk of not realizing the need for explicit coordination, especially in non-routine situations. To avoid this problem, the concept of heedful interrelating has been put forth. It is understood as a form of team interaction that is characterized by the deliberate efforts made by all team members to constantly re-consider the effects of their actions in relation to the goals and actions of others and to the broader context (Weick and Roberts 1993). Presumably, heedful interrelating prevents team members from narrowly following protocols or relying on over-learned responses and primes team members to speak up when necessary.

Given the various problems related to standardization, there is an increasing awareness even in high-risk systems for the necessity to achieve loose coupling by adjusting the balance between stability and flexibility towards more resilience (Hollnagel *et al.* 2006; Weick and Sutcliffe 2001).

4.2 Flexible Routines through Flexible Rules

The newer perspective on routines embodying a duality of principle and practice is thought to be useful for balancing stability and flexibility in organizations in the face of uncertainties. However, much of the research is conceptual and descriptive, rendering it difficult to derive concrete proposals for organizational measures and instruments that might support flexible routines. It is suggested that establishing flexible rules may be an appropriate means to further flexible routines (Grote *et al.* 2009). This suggestion differs from other approaches that argue for few, but quite deterministic rules to provide sufficient orientation in very uncertain environments (*e.g.*, Brown and Eisenhardt 1997).

To further explore this suggestion, a distinction between routines and rules is required as well as definitions of what a flexible routine and a flexible rule might be. Routines are understood as defined by Feldman and Pentland (2003, p. 95) in terms of "repetitive, recognizable patterns of interdependent actions, carried out by multiple actors". While the actual behaviour constitutes the routine in practice, the description and abstract understanding of these behaviour patterns constitutes the routine in principle. Rules, that is, the written-down formal descriptions of standard operating procedures, are artefacts, which are related to and interact with, but are not identical to either routines in principles or routines in practice.

Rules are often written by management with the intention of creating routines in practice by following the rule. In this case, the match between performance and the rule is the measure of management control. Rules might also express the agreement between actors on what to do or the managers' understanding of what their subordinates do. Then the rule is more of an after-the-fact description of routines in practice. One may also look at the relationship between rules and routines in principle: the rule may serve as a formal statement of the routine in principle, but it may also be a public statement about how things ought to be, while everybody in the organization shares another understanding of how things really need to be done.

The concept of flexible routine (Howard-Grenville 2005) is built on the assumption that routines are always enacted, which automatically leads to variation and change in the routine. However, the amount of variation and change needed or allowed and thereby the degree of flexibility, may vary. Therefore, it is useful to distinguish between flexible routines and flexible use of a routine. How much flexibility is encouraged will be at least partially determined by the rules that are attached to the routine. A flexible routine – possibly supported by rather open and flexible rules – incorporates leeway for decision-making. Flexible use of a routine on the other hand, may imply that a rule is adapted by the user without the rule itself explicitly allowing such an adaptation. In this case flexible use is usually considered a violation with associated sanctions.

In Howard-Grenville's (2005) study, the use of a roadmapping routine in a chip manufacturer was analyzed. This routine was in itself sufficiently flexible to allow adaptation to different types of technology development and to different goals by different actors. The rules attached to the use of the routine left open, for instance, how much the roadmap has to focus on equipment or process specifications. Also, entries in the roadmap, which presumably were impossible to change once they were decided upon, could be changed under certain circumstances, without this being considered a violation. Additionally, examples of flexible use of the routine without backing by particular rules were found as different actors worked with the routine to achieve different individual goals. Whether these flexible uses are permitted or sanctioned depends on the truce established around the use of a routine. In Howard-Grenville's example, different uses were accepted because they allowed broader agreement on using the roadmapping routine across different technical departments.

4.2.1 Rule Types

Rules as support for loose coupling have not been researched much to date (Grote 2004a) and systematic research into the design and management of rules, usually focusing on safety-related rules in high risk organizations, has only recently begun (Bax *et al.* 1998; Hale and Swuste 1998; Leplat 1998; Reason *et al.* 1998).

Rasmussen (1997, p. 191, italics in the original) has argued that rather than striving to control behaviour by *fighting deviations* from a particular pre-planned path, the focus should be on the control of behaviour by *making the boundaries explicit and known* and by giving opportunities to develop *coping skills at boundaries*". Rules then would have the function of clarifying boundaries and suggesting ways of handling system states close to those boundaries. In line with this approach to rules, authors have begun to develop typologies of rules in order to help the design of rule systems directly tailored to the needs both for guidance and for autonomy and control arising in different stages of action regulation (*e.g.,* Hale and Swuste 1998; Leplat 1998). Hale and Swuste (1998) distinguish between three kinds of rules:

- rules can concern goals to be achieved (goal rules);
- rules can define the way in which decisions about a course of action must be arrived at (process rules); or
- rules can prescribe concrete actions (action rules).

Goal rules may be very useful in pointing out desirable outcomes and priorities and thereby providing an overall orientation for the actor (*e.g.,* "Flights through areas with known or forecast thunderstorms, severe turbulence or wind shear should be avoided whenever possible"). They target knowledge-based action regulation (Rasmussen 1983), that is, conscious problem solving with high demands on cognitive resources. Goal rules may give rise to the problem that broad responsibilities for goal attainment are assigned without adequate support for reaching these goals. Many rule books contain some very general goal rules that appear to define scapegoats rather than help people achieve the goals. They usually read like "Unforeseen situations may require deviations from the rules, the persons responsible in these situations have to decide and act in ways that guarantee safety".

Process rules are particularly well suited to the design of stable but flexible work processes. These rules provide guidance through heuristics for information search, decision-making, and problem solving (*e.g.,* "In order to complete a re-planning of a flight path, any documented cruise system and all means available may be used, such as flight management systems (where available) and data contained in the respective Aircraft Operation Manuals"). Process rules often aim at rule-based action regulation (Rasmussen 1983) which implies conscious choices between fairly well known action alternatives based on well-known procedures with medium demands on cognitive resources. Process rules sometimes also target knowledge-based regulation, when alternatives are less well known or have to be created by means of the heuristics provided. Process rules may also be meta-rules in the sense that they provide guidance on when rules should be used as coordination mechanism versus when other coordination mechanisms, such as mutual adjustment in a team or leadership, are more appropriate. In the rule book of a railway company we studied, there is the fundamental rule, for instance, that in situations not covered by the rule book, all people involved need to make decisions by mutual adjustment.

Action rules are those rules that for many are the "real" rules or standard operating procedures in the sense of prescribing concrete actions (*e.g.,* "If fixed signals for permitting train departure are available, they are to be used"). They target skill-based action regulation (Rasmussen 1983) which happens with little or no conscious control based on highly practiced action patterns freeing up cognitive resources for other tasks. The more decision latitude is provided in the action rule, the more rule-based action regulation is required. Highly prescriptive and detailed action rules are the rules that people usually have in mind when they criticize the rigidity of standardized work processes, because their main strength is also their main weakness: Action rules provide very detailed guidance for actors, which reduces the demands on knowledge, expertise and resources spent on action regulation and coordination; if the rule fits the situation perfectly, this will usually result in "perfect" action, if it does not, chances of actors realizing that and adapting their behaviour accordingly are severely reduced.

This description of different rules types with their particular strengths and weaknesses indicates that not only the amount of rules, but also the specific qualities of rules will affect the balance of stability and flexibility of work processes. Adler and Borys (1996) described qualities of "good" rules by means of contrasting coercive and enabling bureaucracy. In order to avoid stifling responsiveness and adaptation, which is characteristic of coercive bureaucracy, rules have to facilitate responses to real work contingencies, increase process transparency, and provide contextual information. This creates an enabling type of bureaucracy where rules serve as guidance for individual action and support for smooth coordination.

4.2.2 Effects of Rules on Team Coordination

In their book on the dynamics of rules, March and colleagues (2000) have pointed out the importance of rules for supporting implicit team coordination, that is, coordination based on a shared understanding of the situation and the task which does not require resources being spent on the coordination as such. Few empirical studies have addressed this assumed relationship between standardization and team coordination. In a study by Grote and colleagues (2004), coordination behaviour in cockpit crews and anaesthesia teams as examples of work environments with high versus low standardization was compared, including an analysis of the rules relevant for the observed settings using the Hale and Swuste (1998) categories. As was expected there were generally more rules for the aviation setting and there were also fewer process rules and more action rules compared to the anaesthesia setting. Considering the higher degrees of operational uncertainty contained in handling a patient compared to flying an aircraft, the less specific rules in medicine seemed appropriate. Very rare in both settings were rules that also provided a rationale for the rule. This last finding is interesting in light of a study by Wright and colleagues (1998) in which the annotations in a Quick Reference Handbook by an airline pilot were analyzed. Almost 40 percent of the annotations concerned explanations on why the procedure was the way it was, indicating the need of the users of procedures to understand their reasoning, which can then be used to apply the procedure adaptively. Possible adaptations and support for priority setting were the focus of another 25 percent of the annotations.

Contrary to their original assumptions, Grote and colleagues (2004) found that anaesthesia teams coordinated more implicitly than cockpit crews despite having fewer written rules guiding their behaviour. This finding may be related to more informal rules and the immediacy of common action in a shared visual field for the anaesthesia teams (Hindmarsh and Pilnick 2002, Zala-Mezö *et al.* 2009). For the aviation data, a clear link between higher levels of explicit coordination and higher levels of performance could be established, as has been found in other studies as well (Tschan *et al.* 2006; Waller 1999). This result hints at the importance of backing up standards with a constant effort to reassure a common understanding of the situation and the relevance of the standards for the situation (Weick and Roberts 1993).

Another set of analyses in the study by Grote and colleagues (2004) concerned patterns of coordination within each professional setting, comparing work phases with different degrees of standardization. One important finding here was that high levels of personal leadership in highly standardized situations appeared to be related to worse team performance. Another finding was that cockpit crews performed worse when first officers used higher levels of implicit coordination, pointing to the importance of experience for coordinating action on the basis of shared views of situations and their demands.

Generally, the findings indicated the importance of process rules in helping teams deal better with the demands of adaptive coordination, that is, demands to change coordination mechanisms depending on the nature of the task, the uncertainties to be handled, the task load, and actors' expertise.

4.2.3. Procedures for Defining Appropriate Rules

Besides the rules and procedures themselves, the process of generating and modifying rules is crucial in providing or impeding flexibility. "Formality – government by abstraction – ... [has] to be embedded in a social process that produces accuracy, communicability, and a trajectory for improvement" (Stinchcombe 2001, p. 179). Bourrier (1998) showed in her comparative case studies on the organization of maintenance work in four nuclear power plants how the maintenance workers' influence on the writing and modifying of procedures was positively related to them following the ensuing rules. Similarly, Adler and colleagues (1999) described the extensive standardization and formalization of routine operations in the NUMMI Toyota plant, which were the result of highly participative definition and refinement of standards. "Departures from procedures were typically treated as opportunities for learning rather than as threats to authority. [...] While lower levels did not have much autonomy to make decisions without prior consultation with superiors, this apparent centralization usually took the form of 'fact-based' dialogue based on expertise rather than command-and-control domination based on positional authority" (Adler *et al.* 1999, p. 54). These examples can be taken as evidence for the importance of higher order autonomy (Grote 2004; Klein 1991), that is, autonomy in restricting one's own operative autonomy for effective performance in highly standardized work settings.

Bax and colleagues (1998) looked both at the type of rules (formal versus informal) and the ways rules could be changed which they called the regulation regime. They found that perceived legitimacy of formal rules was linked to the regulation regime, but interestingly, rule enforcement and consultation about rules was positively related to perceived legitimacy, while participation showed a negative relationship. This result may be understood in terms of what Nelson and Winter (1982) have called the truce function of rules, which may require a particularly clear cut management responsibility for rule formulation and enforcement.

4.3 A Field Study: Rules Management in a Railway Company

A concrete example for evaluating rules and their effects on managing uncertainties has been provided by Grote and colleagues (Grote 2008b; Grote *et al.* 2009). Their study was conducted in a railway company, a setting considered to be particularly suited for investigating rules and routines as railways are among the most proceduralized organizations. Railway organizations have a very long history of using rule books as the basis for coordination mainly due to their highly distributed nature of operation. Furthermore, until very recently no means for direct communication between the different actors, such as train drivers, controllers and maintenance workers, existed, necessitating impersonal forms of coordination through standards and programmes (Hale and Heijer 2006). At the same time, rule violations and inadequacies of rules and rule management have been found to be crucial to safety problems in rail organizations (Farrington-Darby *et al.* 2005; Hale *et al.* 2003 ; Lawton 1998).

The general purpose of the study was to provide an analysis of existing rules and of the rule management process shared between the state regulator and the railway organization. From a research perspective, the adequacy of existing rules and rule management processes for providing optimal support for the management of uncertainties in rail operations was focused upon. For the railway company and the regulator, the study was to provide a basis for reflecting upon four themes: adequate detail of rules, interplay of rules and training, interplay between rule design and rule auditing, and distribution of responsibility between actors involved in rule design, rule following, and rule monitoring.

Data were collected via rules analysis of a sample of rules based on two work processes (train departure, coupling and braking of cars during shunting operations), interviews with executives responsible for rule management and with operators, observations of the relevant work processes, analysis of documents on rule management, and analysis of incident reports.

For the rules analysis, categories suggested by Hale and Swuste (1998) were used in a slightly modified form:

1. Level of action regulation
   - Goal (the rule only defines the goal without determining the way of achieving the goal)
   - Process (the rule describes the processes needed to determine the right course of action)
   - Action (the rule prescribes a concrete operative course of action)
2. Obligation (advice vs. command)
3. Decision latitude (with vs. without decision latitude)
4. Distribution of responsibility (responsible person(s) mentioned vs. not mentioned)
5. Exceptions (with vs. without exceptions)
6. Reasons (with vs. without reasons)

Table 4.1 presents the results of the analysis of rules concerning the coupling and braking of cars during shunting operations and the moving of trains. Regarding the moving of trains, train departure was focused upon, for which new rules had recently been devised by the regulator. In the analysis, the old and new rules for train departure were compared in order to obtain a reference profile for the changes made by the regulator.

Table 4.1. Results of the rules analysis (adapted from Grote *et al.* 2009)

Rule type	Shunting		Moving trains (New rules related to train departure)		Moving trains (Old rules related to train departure)	
	Frequency	Percent	Frequency	Percent	Frequency	Percent
Total no. of rules	28	100	32	100	38	100
Goal rule	0	0	1	3	0	0
Process rule	6	21	9	28	2	5
Action rule	22	79	22	69	36	95
Advice	0	0	1	3	0	0
With decision latitude	20	71	15	47	14	37
With responsible person(s) mentioned	2	7	17	53	13	34
With exceptions	7	25	6	19	7	18
With reasons	1	4	0	0	1	3

Table 4.1 shows that very few goal rules existed, which could be regarded positively because goal rules by their very nature do not offer much guidance for the actors on how to pursue the goal. On the other hand, goal rules may be helpful by providing information on system boundaries (Rasmussen 1997), on the meaning behind other more specific rules, and on priorities in certain situations, thereby supporting the actors in their individual priority setting (Wright *et al.* 1998). This may concern very basic priorities, such as punctuality versus safety, or more specific priorities, such as the timing of one's own actions and communication with other actors about these actions.

Compared to the old rules for moving trains, the new rules included more process rules and also a clearer distribution of responsibility. The latter especially had been a declared goal of the rule revision in order to avoid diffusion of responsibility. At the same time, discussions about the stricter definition of responsibility with the employees affected indicated the potential problem of too little sharing of responsibility, thereby reducing information flow and cross-checking between the various actors. This problem is assumed to gain in importance with different actors like train drivers, maintenance crews, shunters, and signallers belonging to different companies within a holding structure and with fewer and fewer of them having had the experience of belonging to the one previously existing company.

The higher level of process rules in the new rules on train departure can be considered a positive development as process rules provide guidance without restricting flexible action. Signallers and train drivers are well qualified people with high professional ethics, who still have to act largely on very defined and prescriptive action rules, "stretching" these, however, as needed, in order to accommodate non-routine situations. So an even higher proportion of process rules might be appropriate.

The high percentage of process rules and even more so, of rules with decision latitude was evaluated more critically with respect to shunting operations. The prevalence of more open and less prescriptive rules on how to handle shunting operations has to be seen in the context of the shunters' comparatively low qualification level and their reluctance to take responsibility for using the decision latitude offered. The executives responsible for rule making pointed out that shunting occurs in so many different contexts (within stations, on shunting yards with varying degrees of automation, etc.) that rules have to be less specific. So the issue becomes whether or not the current level of qualification and training is sufficient for handling this flexibility. Efforts are underway in the railway company to improve training for shunting accordingly.

One may also raise the issue, brought up in other rail organizations as well (Hale and Heijer 2006), as to whether rules for shunting are paid less attention to by rule makers because they mainly concern the safety of the individual workers, rather than the safety of passengers, which is more political and prevalent. This is also related to the basic issue of rules having different purposes for different actors in an organization (Leplat 1998; Power 2004). While management sees devising rules and enforcing rule compliance mainly as a way to fulfil their responsibility of controlling employees' actions, employees evaluate rules in terms of their potential for supporting or hindering adequate action. Less risk implies less management attention and therefore fewer and less specific rules. On the other hand, from the shunters' point of view, their low level of education and training creates more demands for guidance through specific rules.

A high percentage of rules with exceptions was found, which can mainly be explained by the fact that a variety of local conditions has to be covered, *e.g.*, regarding the level of automation, the number and distribution of personnel (trains with/without conductors, stations with/without signallers etc.), and the particular geographical layout. The main issue to be discussed with respect to these exceptions is the level of detail the rules should have and whether local knowledge should be incorporated in the rules or rather be provided by complementary training. The fundamental dilemma of not being able to create prescriptions that are, at the same time, general, simple, and detailed became obvious here.

Finally, rules very rarely included reasons, raising the question of whether training is designed to provide the necessary background information. Given an increasing pressure for interoperability of rail systems, requiring personnel with more and more diverse educational and training backgrounds having to work together, rules might have to include more explanations in the future.

4.3.1 Implications for Developing Flexible Routines through Flexible Rules

Findings from the field study clearly show difficulties in aligning rules with given uncertainties and the actors' competencies for handling uncertainties. Two opposite patterns of mismatch were observed: The less qualified personnel had to operate with rules that provided more decision latitude, while the more qualified personnel was confronted with very detailed and restrictive rules. For the case studied, the identification of these mismatches led to the following conclusions: (1) Uncertainties in shunting operations need to be reduced through more detailed and prescriptive rules and/or shunters' individual competencies need to be increased to use adequately the provided decision latitude; (2) signallers and train drivers need to be supported more in coping with uncertainties, potentially through more process rules, including rules that support adaptive coordination, *e.g.*, switching from proceduralized, implicitly coordinated action to explicit coordination through direct communication and mutual adjustment between different actors. However, in providing signallers with more scope of action, a critical balance with the decision latitude for other actors has to be maintained. The analysis of the rules and even more so, the analysis of some incident reports, seemed to indicate that the signallers sometimes handle uncertainties in a way that reduces other actors' possibilities for handling their part of the situation appropriately, *e.g.*, by providing not enough or untimely information about their actions or the actions required by the others (compare also the rail example described in Chapters 2 and 3). A power differential between different professional groups is apparent, as signallers have the power to shift uncertainties to other actors who sometimes do not have the adequate resources to handle these uncertainties (Hale and Heijer. 2006; Marris 1996).

By reinterpreting these findings through an organizational routines lens, more dangers potentially arising from these mismatches become visible. In the case of shunters, the lack of regulation could lead to the development of routines disconnected from other parts of the organization and not in line with organizational safety standards. If no other forms of coordination, like shared cultural norms on safe operations and frequent communication with other parts of the organization, take place, this integration deficit could compromise the organization's safety. In the case of signallers, where the workflow and coordination was found to be well (if not over-) regulated, this high level of standardization could suppress adaptation of routines when the actual state of affairs differs from the one assumed in the regulation, thus compromising safety as well.

4.4 Systematic Rule Management as Support for Creating Flexible Routines

For integrating the different considerations presented in the previous sections, the framework for designing and evaluating rule management *systems* (see Fig. 4.1) developed by Hale and colleagues (2003) is very useful. They distinguish seven steps from analysis of work processes and coordinaton requirements, definition of work processes that require rules, and the development, writing, training and executing of those rules to monitoring, enforcing, evaluating and modifying the rules. The first three steps are of particular importance as regards the integration of rule management with uncertainty management and the aim of creating flexible routines.

Step 1 and 2 in the framework by Hale and colleagues (2003), analysis of work processes, coordination requirements and appropriateness of standardization, can be covered by analyses following the decision process described in Chapter 3. Costs and benefits regarding reducing, maintaining or increasing uncertainties through rules in particular need to be focused on. Also, belief systems regarding the usefulness and effects of rules are to be explored. Taking the move from prescriptive regulation to goal-oriented legislation as an example (Kirwan *et al.* 2002), one can argue that prescriptive regulation in combination with the necessary control instruments reduces uncertainty at a high cost, which in recent years has been regarded as too high. Goal-oriented legislation, on the other hand, leaves the control of uncertainty to the organization and reduces costs for surveillance by the regulator, but maintains or even increases uncertainty, because more leeway for operations is provided. In this shift of regulatory practice one can also find changing belief systems, where today self-regulation and decentralized responsibility is stressed more compared to centralized authority and accountability. The current crisis in the financial system has led to major doubt regarding the appropriateness of this new regulatory approach, though.

Step 3, the development of applicable rules, should take the different rule types and their advantages and disadvantages as a starting point. A general guideline is

- to use goal rules in combination with process rules especially in situations where uncertainty is increased,
- to use process rules when uncertainty needs to be maintained, and
- to use action rules when uncertainty can be reasonably reduced.

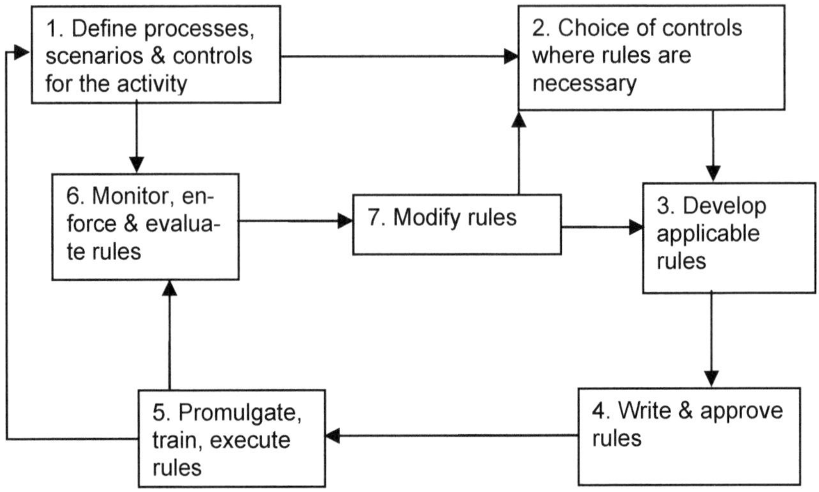

Figure 4.1. The rule management process (from Hale *et al.* 2003)

As the enactment of organizational routines inevitably allows for variation and change, decision latitude for the rule user should always be explicitly designed into rules, thereby channelling the adaptation of rules into routines in practice. By defining boundary conditions for rule use more systematically, flexible rules do not undermine safe performance (Rasmussen 1997).

Additionally, the process through which rules are defined and managed has to be determined, especially with respect to participation of rule users and the level of the organization at which rules should be decided upon. Hale and Swuste (1998) have suggested some criteria to help decide at which level of the organization action rules in particular should be defined:

- predictability of the system – the higher the predictability, the more action rules can be decided upon at higher levels of the organization;
- innovation rate in the system – the higher the innovation rate, the more action rules need to be decided upon at operative levels of the organization;
- interaction requirements – the higher the interaction requirements, the more action rules need to be decided upon at higher levels of the organization;
- local expertise – the higher the local expertise the more action rules should be decided upon at operative levels of the organization.

These criteria can be easily related to the general issue of minimizing versus coping with uncertainties, where minimizing uncertainties through centrally determined action rules is only recommended in systems with a generally low level of uncertainties (high predictability, low innovation rate). Also, these recommendations are linked to the idea of higher order autonomy stating that people should be involved in decisions restraining their operative autonomy (Grote 1997; Leplat 1998), especially in organizations with high local expertise.

One way to continuously monitor and improve rules is informal "toolbox" meetings which permit open discussion of workers' experience with certain rules, possibly also rule violations. Such meetings keep rules alive and allow for adaptation of rules in line with changing work demands.

Some basic issues in rule management remain very difficult to resolve, above all, the contradictory requirements of reducing demands on cognitive and organizational resources during non-routine, time-critical situations while also increasing response flexibility and adaptability, which requires extra resources. Rules and routines generally help reduce demands on resources, but the more open they are, the more they lose this advantage. It was recommended earlier that goal and process rules should be used in situations with high uncertainty. Both of these pose high demands on decision-makers, though, which in such situations in particular they may find very difficult to meet.

Current discussions around the computerized Traffic Alert and Collision Avoidance System (TCAS) in aviation illustrate this dilemma. These discussions were fuelled by the tragic accident over Überlingen in Southern Germany around midnight on the 1st of July 2002, where two aircraft collided in an otherwise empty sky (*e.g.,* Weyer 2006). Many problems led to this accident, but the final cause was that one cockpit crew followed the order by the TCAS to descend, while the other crew followed the order by air traffic control (ATC) in Zurich to descend contrary to the TCAS order to climb. This was possible because the two crews came from different airlines and different countries that had different rules in operation. One crew was from a US airline where the rule is to give TCAS priority over ATC orders, while the other was from a Russian airline where ATC has priority over TCAS. Following the accident, an international standard was issued stating that TCAS always has priority over ATC. However, final responsibility stays with the pilots, which leaves them potentially with having to make decisions within seconds to override TCAS without much further guidance provided by the rules. In the flight operations manual of a European airline this reads as follows: "Follow resolution advisory (RA), (*i.e.*, the actual order given by TCAS) immediately unless considered unsafe or the conflicting traffic is identified visually and you decide that no deviation from the current flight path is needed". During a recent workshop on the future of air traffic management in which I participated, a test pilot from Airbus commented on this problem by stating that pilots are very well paid to be made responsible even in situations where they have no control. This point will be discussed again in Chapter 6 with respect to designing automated systems.

4.5 Chapter Summary

In this chapter, coordination in organizations was discussed with the specific focus on rules and routines as a coordination mechanism. From a contingency perspective, the issue is to establish a balance between stability and flexibility by rules and routines that allow uncertainty to be coped with while at the same time, providing sufficient orientation to ease coordination demands. Based on newer research on organizational routines, it was argued that routines can be sufficiently flexible to support adaptation in unforeseen situations, especially when they are backed by flexible rules. Subsequently, some research on different types of rules and their relationship to coordination patterns and performance was discussed, indicating the importance of a more sophisticated perspective on rules and standardization. Instead of only taking the amount of rules as an indication of more or less standardization, types of rules need to be studied. One way of distinguishing rule types is based on the level of action regulation targeted; that is, goal, process, or concrete action. An example of a rules analysis was presented which showed a mismatch between rule types, uncertainties to be handled and actors' competence profiles. Such analyses can serve as input into the redesign of rules aimed at creating flexible routines capable of absorbing uncertainty when needed. As part of a systematic rule management, different ways of managing uncertainties by means of different types of rules need to be explored, for which the decision process presented in Chapter 3 can provide some guidance. Furthermore, the process used for (re)designing rules has to be chosen, considering the participation of future rule users and the level of the organization at which decisions on rules are made. Finally, it was acknowledged that rule management will only be able to partially resolve the dilemma stemming from the contradictory requirements in non-routine situations of reducing demands on cognitive and organizational resources, while also increasing more resource-intensive behavioural flexibility and adaptability.

5

Managing Uncertainties through Collaborative Planning

Co-authored by Hannes Günter

"It is no accident that much of the literature on the management or administration of complex organizations centres on the concepts of planning or controlling. Nor is it an accident that such views are dismissed by those using the open-system strategy." (Thompson 1967, p. 6)

Planning aims to turn uncertainties into predictable action paths and outcomes. The more uncertainties there are, usually also in combination with multiple interdependencies within and across organizations, the more planning is usually undertaken. Uncertainties may concern lack of knowledge or ambiguities regarding the desired outcomes, the current situation, or things to be learned from past states. New information technologies are embraced by many companies as a means to support planning and ultimately control by increasing transparency and predictability of organizational processes and events. Whether more planning really increases control is debatable and there is even some evidence that a lot of the time it does not. Particularly, uncertainties about goals to be achieved, possibly also resulting from conflicting goals, and about what to learn from the past, are difficult to reduce by planning. In this chapter different approaches to planning and their success in achieving control are presented. The focus in this discussion is on operational – not strategic – planning as it relates to the execution of work processes. Subsequently, a model of collaborative planning is suggested and findings are presented concerning its validity in the context of intra- and interorganizational supply relationships. Finally, the intricate relationships between uncertainty, planning, autonomy and interdependence in organizational networks and their consequences for organizational planning are discussed and illustrated by a case study.

5.1 Approaches to Planning in Organizations

Operational planning in organizations is usually discussed from the perspective of internal logistics, especially linking demand forecasts, purchasing, and production planning and scheduling, and from a broader interorganizational perspective within the framework of supply chain management (SCM). SCM involves "all activities associated with the flow and transformation of goods from the raw materials stage (extraction), through to the end user, as well as the associated information flows" (Handfield and Nichols 1999, p. 2). SCM concerns procurement of raw materials or parts needed for production of a particular product as well as the logistics involved in getting products to customers. Depending on the chosen degree of specialization across organizations, SCM may require coordinated action of a multitude of actors spread across a multitude of organizations.

5.1.1 Feedforward versus Feedback Control

The two approaches to uncertainty management discussed in Chapter 3, minimizing versus coping with uncertainties, differ particularly with respect to the chosen perspective on planning. Planning can be regarded as a form of feedforward coordination, which prescribes actions based on a thorough understanding of task requirements and task interdependencies (Thompson 1967; Van de Ven *et al.* 1976). If people then behave according to the planned courses of action, coordination is achieved without much need for explicit coordination by supervisors or mutual adustment between the actors involved. This view lies at the heart of the minimizing uncertainties approach, which follows a rationalistic model of human action (*e.g.*, Miller *et al.* 1960), where plans are seen as blueprints for action. This approach is exemplified by the claim that plans are analogous to computer programmes. Just as computer programmes control the execution of various processes, plans are thought to control human behaviour. Planning is considered to be hierarchically organized, in terms of an orderly progression from long- to medium- and short-term planning and from more general to increasingly more specific plans. This is exemplified in the difference between production planning and scheduling in many organizations, where planning is concerned with the more long-term and general framework for production and scheduling provides short-term and very specifically prescribed action paths.

A contrasting perspective on planning is based on a theoretical framework stressing the situatedness of human action. Here the focus is on heterarchical processes involved in attempting to achieve various super- and subordinate goals at the same time. Action is regarded as opportunistic in the sense that options are drawn up, elaborated upon and chosen flexibly based on characteristics of evolving situations. Plans help to adapt action opportunistically and thus serve as a resource for situated action. "The function of abstract representations is not to serve as specifications for the local interactions, but rather to orient or position us in a way that will allow us, through local interactions, to exploit some contingencies of our environment and to avoid others" (Suchman 1987, p. 188). Plans, in other words, serve as vehicles to create "prepared minds" (Kaplan and Beinhocker 2003). Coordination between actors happens through mechanisms that allow feedback control, such as mutual adjustment and personal leadership. This perspective is core to the coping with uncertainties approach.

Empirical evidence generally supports this second perspective, especially for planning in turbulent environments (Mumford *et al.* 2002; Naveh 2007). Regarding long-term buyer-supplier relationships, Mesquita and colleagues (2008) found that reliance on relational governance, which concerns behavioural norms of open information exchange, flexibility, and solidarity, was particularly important for highly complex products and production processes, presumably also involving more uncertainty. Reliance on contract completeness, that is coverage of all imaginable contingencies in contracts, was more important when dependence on the particular buyer-supplier and therefore, the needs to safeguard the relationship, were high. In the present context, these findings can be taken to indicate that more open and opportunistic planning processes are required with higher levels of uncertainty. In experiments with students working on a business game, Mathieu and Schulze (2006) showed that in well-performing teams deliberate formal plans served as a framework for more dynamic and episodic planning. In a study of product development teams, Eisenhardt and Tabrizi (1995) found that teams following an experiential project management strategy, comprising multiple design iteration, extensive testing, and frequent milestones outperformed teams that focused on detailed planning and rewarding schedule attainment. Project management at Google may serve as a good example of the experiential strategy, with detailed quarterly plans embedded in rather general roadmaps and intensive communication and coordination within development teams as well as across a broad range of domain experts. Google's success seems to indicate that the enormous coordination costs of such a strategy are outweighed by its benefits.

5.1.2 Tightening versus Loosening of Interdependencies

For both intra- and interorganizational planning, basic difficulties concern the coordination of highly interdependent processes across organizational units and whole organizations in relation to partially incompatible goals. In intra-organizational planning this may concern conflicts between goals such as generating new customer orders, keeping materials stock low, and shortening lead times. In inter-organizational planning similar conflicts may arise, spread across different companies, that add conflict potential, as costs and benefits of goal attainment may be distributed unequally between the companies involved. A crucial issue for handling such conflicts is the distribution of autonomy, control and power among the different partners and the ensuing willingness or reluctance to cooperate. It is generally assumed that cooperation is most likely between partners that are mutually dependent – even though the supporting evidence is equivocal, as will be discussed further in later sections of this chapter. Based on this assumption, many concepts for improving intra- and interorganizational planning imply tightening interdependencies between the partners involved.

In intra-organizational planning, specifically in production planning and scheduling, concepts such as mutual constraining (McKay 1992), perspective taking and interlinking (Zölch 2001) and shared autonomy between primary work systems responsible for production and secondary work system responsible for planning (Wäfler 2001, 2002) have been suggested. Jackson and her colleagues (2004) have provided very interesting case study material on how production schedulers in manufacturing plants build personal networks that rely on the reciprocity of small favours in order to handle conflicts arising from interdependencies.

The common idea behind these different concepts is a general framework for cooperative planning based on a shared understanding of interdependencies and willingness to employ one's own autonomy and means of control in ways that help to manage these interdependencies in the interests of all parties involved (*e.g.*, Barratt 2004; Mitchell and Nault 2007; Ring and Van de Ven 1992). From the research by Wäfler (2001), Windischer (2003; Windischer and Grote 2003) and Gauthereau and Hollnagel (2005) it can be argued that local control for implementing plans should always be maximized, while autonomy in planning and implementation should be distributed according to task interdependencies and goal agreement. When local actors can identify with the goals underlying a plan even though they have not participated in defining the goals or the plan to achieve them, for instance, because there are shared interests across all parties involved, autonomy can stay with a central decision-maker. Also, when task interdependencies are sequential in nature, autonomy has to remain with a central planning body.

In supply chain management (SCM) research concerned with interorganizational planning, much attention has been devoted to strategic networks as an effective organizational form for managing interdependencies. Strategic networks are organized around a "hub firm" which has had a central position in setting up the network and which is also central in coordinating the network's activities (Jarillo 1988). Strategic networks are considered to be effective when tasks are shared between the partners in such a way that everybody can focus on their core competence and when uncertainties are distributed according to capabilities for coping with them. This implies in particular, that the hub firm does not transfer all uncertainties to the weaker partners in the network. As Scott (1981) has put it, the aim is to balance autonomy-related losses with interdependence-related gains through risk-sharing.

Despite the presumed advantages of supply chain integration, surveys have found that implementation of SCM is deficient in most companies (Bagchi *et al.* 2005) and not necessarily linked to better performance (Birdi *et al.* 2008). It is generally assumed that inadequately balancing autonomy and interdependence lies at the heart of the difficulties encountered in managing supply chains (*e.g.*, Scott 1981; Sydow and Windeler 1998). Relying on supply chain partners to provide crucial functions for one's own business decreases autonomy and will usually only be accepted if the loss in autonomy is compensated for by major economic gains. However, interorganizational interactions may also be a basis for stability if uncertainties can be managed by stable patterns of collective association (Astley and Van de Ven 1983). Scott (1981) has listed a number of strategies companies can use to proactively handle interdependencies in the relationship with other companies, such as building internal buffers and thus reducing uncertainties related to the relationship, or bridging the uncertainty by sharing the risks *e.g.*, through contracting or joint ventures. In their analysis of the supplier relations at the Toyota plant NUMMI, Adler and colleagues (1999) found evidence for this positive approach to interdependence, *e.g.*, by NUMMI relocating some innovation tasks to the supplier firms where the required knowledge was available based on a shared standard master schedule.

Tension between autonomy and interdependence is strongly influenced by the overall amount of uncertainty with which the supply chain is confronted and the resulting requirements for flexibility (Stevenson and Spring 2007). The difficulty in interorganizational planning from this perspective again lies in contradictory requirements (Nassimbeni 1998). On the one hand, coordination mechanisms should not be too rigid in strictly binding the nodes, so that flexibility in the system is not lost. On the other hand, the activities of each system node should be synchronized with the activities of the whole system. Furthermore, a systematic analysis of uncertainties and decisions on managing uncertainties as suggested in Chapters 2 and 3 become far more complex when several organizations with often very diverse contexts have to be considered. One possibility is to carry out such analyses in pairs, looking at inter-linkages and the particular contexts of two companies at a time, and to try to draw up an overall picture from these separate analyses. Pairs could be determined, for instance, by studying the strengths of relationships first and choosing those companies with the strongest links (Günter 2007). Given the enormous complexity of such analyses, the more generic solution put forth by many authors is to further collaborative planning among all partners, so as to permit handling uncertainties cooperatively (Barratt 2004; Danese 2007; Lorenzoni and Lipparini 1999).

A final issue which impedes effective management of interdependencies is the unequal distribution of power, specifically the power to handle and to transfer uncertainties (Marris 1996). Marris' impressive account of the importance of framing power in terms of transfer of uncertainty mainly builds on examples from social, particularly community, planning. For instance, large companies may demand services from communities, such as permits for land usage, without reciprocating commitments and possibly even without agreeing to the deal in the end. But any supply chain organization or the relationship between sales and purchasing involves similar mechanisms. "Plans [...] represent at once collaborative strategies and ideal rationalizations of the disjointed interests out of which a set of social and physical relationships have been constituted. The ideological rationalization, because it obscures and often ignores the realities of power, tends to undermine the collaboration it intends to promote" (Marris 1996, p. 142). Marris argues for furthering collaborative strategies because of the reduction in overall uncertainty and the increase in collective control which it offers. However, "collaborative strategies, for all their advantages, depend upon patience, mutual understanding, farsightedness – the very qualities of relationships that are most often lacking in conditions of uncertainty" (Marris 1996, p. 148). Despite potential long-term advantages of collaboration, actors short-sightedly perceive their resources to be sufficient to remain autonomous and therefore do not enter into collaborative agreements. And even the weaker actors might shrink from collaboration because their autonomy is not sufficient to sustain their commitments. On a more optimistic note, Marris argues in summation that it is possible to show sufficient benefits from collaborative management of uncertainty so that trust is built up as a bridging strategy for the remaining uncertainties.

Similar arguments have been put forth by Jarillo (1988) and many others after him for overcoming power gradients in organizational networks (*e.g.,* Danese 2007; Ring and Van de Ven 1992, 1994; Stevenson and Spring 2007). Often the talk about the importance of collaboration and trust remains on a rather superficial level, though, without much conceptual or practical development to help introduce these concepts into the actual management of supply relationships. This shortcoming is addressed in the following sections, where a model of collaborative planning is introduced and evidence for its usefulness in intra- and inter-organizational planning presented.

5.2 A Model of Collaborative Planning

Collaborative planning entails deliberate efforts by individuals to coordinate their planning processes with others. Individuals predominantly plan according to the goals of their own reference system, for instance, their organizational unit or their company, but in cooperative relationships they also try to further joint optimization of plans by taking into account other individuals' reference systems. Thus, collaborative planning includes aspects that enable the other person to recognize how individual plans should be adapted, according to what criteria the planning can be optimized, and which restrictions in the common planning must be accepted. In short, collaborative planning is about how well individuals are able to orient their plans towards each other to reach a joint optimization of the planning process across departmental and organizational boundaries. In line with this argument, research focused on the individual cognitive tasks of schedulers has shown the importance of their social or networking role (Crawford *et al.* 1999; Jackson *et al.* 2004; McKay *et al.* 1989; Sanderson 1989).

Psychological theory usually is concerned with individual planning (*e.g.,* Funke and Glodowski 1990; Hoc 1988). According to action regulation theory, an individual's actions are seen as units of work activity, which are organized hierarchically and sequentially. Before one goal is attained, the related sub-goal must be achieved, at which point the agent can progress to the next goal (von Cranach and Tschan 2001). A prototype of a regulation cycle consists of goal determination, orientation, development of an action programme to reach the goal, decision about the execution form, as well as action execution, and finally monitoring of the action and final evaluation of the results. Requirements of the task which the decision-maker has to fulfil, influence action regulation. Planners have to elaborate and adjust their plans more when the task requires the development of a common planning strategy, compared to repetitive tasks which only require information exchange. For example, a planner with repetitive scheduling tasks might not have to put much effort into the development of an action programme in contrast to a planner who has to deal with rapidly changing demands.

Some researchers (Oesterreich and Resch 1985; Resch 1988; Tschan 1995; Zölch 2001) have found regulation cycles to be a characteristic of collaborative planning processes also. Individuals must co-ordinate their actions in relation to the various steps in the regulation cycle. Thus, collaborating individuals must communicate to enable mutual adaptation of the regulation cycles. Resch (1988) asserted that during the establishment of plans for others, the individual must know about constraints in the action field of the other person, also called reference field. In cooperative situations individuals should therefore undertake actions of adjustment like creation of prerequisites for and establishment, monitoring and optimization of the interweaving of actions (Zölch 2001). For the model of collaborative planning presented here, it was assumed that regulation cycles of collaborating individuals are overlapping at some points and that individuals have to relate their actions to each other at different stages of their planning activities.

As mentioned earlier, the assumption of a hierarchical structure of action has been criticized (*e.g.,* Suchman 1987). Even representatives of the action regulation approach admit that real-life activity is opportunistic and characterized by "muddling through with some planned episodes" (Hacker 2001, p.58; Volpert 1994). The view of planning as a process is highlighted in activity theory. Planning is seen as being embedded in the social and cultural background (Das *et al.* 1996; Leontjew 1981; Rogoff *et al.* 1994; Scholnick and Friedman 1987). This approach considers revisions as an element of planning. As people are interacting with their environment, the individual's decisions and observations in the planning process suggest opportunities for plan development (Hayes-Roth and Hayes-Roth 1979). Plans therefore serve as an orientation and must support flexibility of human actions (Suchman 1987). Flexibility is required especially for collaborative planning (Rogoff *et al.* 1987), because active and dynamic contributions from various individuals embedded in potentially different social traditions have to be integrated.

Different characteristics support adaptive processes in planning according to contextual requirements. Firstly, explicit planning is necessary, allowing all persons involved to be informed about the actions of the other persons and to obtain a regular update for the mutual adjustment of the planning process. This concerns the communication of anticipated events, allowing the co-actors to proactively adapt their planning. Hence, the communication of modifications in the planning process is also essential. Secondly, the level of detail of the planning has to be adequate to allow the other person to modify the plan. Advance planning and improvisation should be combined, that is, skeleton plans have to be elaborated to various degrees during action. Planning data should give the other person(s) enough freedom to re-orient the planning with respect to the existing resources and planning constraints.

To create plans successfully, it is necessary for the individual to recognize the sequential order of events, relevant boundary conditions, intermediate goals, the availability of alternatives, and the adequacy of the plan's level of detail. To successfully execute plans, it is necessary to be able to monitor planned action, diagnose errors, revise plans, and possibly also cancel plans and switch to alternative courses of action. These characteristics can serve as a guideline for deriving characteristics of collaborative planning activities (Windischer 2003; Windischer and Grote 2003; Windischer *et al.* 2009).

Consequently, the model of collaborative planning suggested by Windischer (see Figure 5.1 and Table 5.1) distinguishes between the creation and execution of plans. During plan creation, knowledge about conditions in the planning environment of the other actors is necessary for collaborative planning. Whereas the individual must be able to recognize the sequential order of the events, cooperation requires anticipated events to be communicated by each person. For example, the transfer of forecasts can enable the supplier to foresee the provision of capacities for future demand. Furthermore, in collaborative planning – besides the recognition of time-, material- and person-related prerequisites of the individual actions – knowledge about the reference action field must be provided. For example, if constraints in the production planning of the other company are known, it becomes obvious for the other person why specific orders often cause problems and how the plan must be adapted.

Another important aspect for collaborative plan creation is the formulation of lateral agreements. Each individual establishes intermediate goals, which, during the collaborative planning process, have to be reformulated as common goals. In interorganizational relationships, the agreement on "frozen zones", that is certain time frames during which the modification of planning data is not permitted, leads to a higher planning stability. Furthermore, the availability of alternatives must be recognized in individual planning. In collaborative relationships, persons must agree on alternatives. For instance, problems in attaining the delivery date can be overcome by the replacement of specific components. The individual must decide on the adequate detail of plans in light of the possible occurrence of unanticipated events. In collaborative planning, the adequacy of the common plan must be examined to see how much of the individuals' regulation possibilities for handling unforeseen events are restricted by the common plan.

During plan execution, the monitoring and revising of the plan is crucial. In both individual and collaborative planning, the execution of the planned action must be monitored. For example, information about order status can serve as the comparison between the planned and realized actions. If problems during the execution of the plan occur, the plan must be revised and possibly replaced. Collaborative planning requires that the actors communicate plan modifications to each other, jointly reflect on the appropriateness of the plan, and decide on cancelling the plan if needed. These processes can be supported substantially by simple measures such as regular common meetings.

In Figure 5.1, the different characteristics of collaborative planning as suggested by Windischer (2003; Windischer and Grote 2003) are presented in the middle column and related to the characteristics of individual planning (Funke and Glodowski 1990; in the left and right columns of the model in Figure 5.1). In Table 5.1, the characteristics are described in more detail and examples are given for each. It is assumed that if these characteristics are managed well in planning processes, better performance should result. This assumption builds on the more general view that investments in collaborative planning pay off. Collaborative planning in concurrent engineering, for instance, has been shown to decrease the magnitude of costly rework, both upstream and downstream (Adler 1995; Mitchell and Nault 2007). It is also assumed that joint optimization of planning processes is necessary, especially if the level of uncertainty requires continuous coordination (Ring and Van de Ven 1992, 1994). Two studies that tested these assumptions using the suggested model for collaborative planning will be presented in the next section.

An underlying theme across all characteristics is the handling of uncertain information, which has been found to be a crucial concern in collaborative planning. Loch and Terwiesch (2005) have recently presented principles for exchanging preliminary information that resemble some of the characteristics in the model suggested by Windischer.

The principles most relevant in the present context are the following:

- Don´t treat preliminary information as final, but communicate it at an appropriate level of precision and stability.
- When addressing multiple outcomes is cheap, hedge.
- Updating information is potentially worth a lot.
- Acting upon preliminary information rests on mutual understanding and relationships.

The suggested characteristics in the model of collaborative planning also resemble the notion of relational coordination (Gittell 2000). This concept does not specifically focus on planning, but states more generally that successful coordination is achieved by frequent and timely communication, problem-solving communication, helpfulness, shared goals and knowledge, and mutual respect. In the context of coordination in health care, relational coordination has been shown to be linked to performance in terms of patient outcomes.

	Individual A		***Collaborative Planning***		*Individual B*
Plan creation	Recognize the sequential order of events	→	Communication of anticipated events	←	Recognize the sequential order of events
	Recognize the boundary conditions	→	Knowledge of reference field characteristics	←	Recognize the boundary conditions
	Recognize intermediate goals	→	Lateral goal agreement	←	Recognize intermediate goals
	Recognize the availability of alternatives	→	Negotiation of alternatives	←	Recognize the availability of alternatives
	Recognize the adequacy of the plan`s degree of detail	→	Recognition of planning adequacy	←	Recognize the adequacy of the plan`s degree of detail
Plan execution	Monitor planned action and diagnose errors	→	Monitoring and diagnosis of errors in the common planning	←	Monitor planned action and diagnose errors
	Revise plans	→	Coordination of opportunistic planning	←	Revise plans
	Cancel plan	→	Common reflection/ decision for plan cancellation	←	Cancel plan

Figure 5.1. Model for collaborative planning: characteristics of individual and collaborative planning activities (adapted from Windischer 2003)

Table 5.1. Characteristics of collaborative planning (from Windischer *et al.* 2009)

Characteristic	Definition	Example from demand planning
Communication of anticipated events	Explicit mentioning of expected events; communication of uncertain information regarding probability of occurrence of events	Early warnings regarding anticipated delivery problems; communication of expected, but uncertain orders
Knowledge of reference field characteristics	Exchanging information on conditions in own and other`s field of action; trying out actions	Providing background information on delivery problems; providing information on own capacities, time management, etc.
Goal agreements	Common definition of goals and reciprocal commitment to achieving the goals	Agreements on safety margins; agreements on frozen zones for demand changes
Negotiation of alternatives	Agreement on a deviation from the original plan	Agreement on express delivery; agreement on taking back unnecessary material
Recognition of planning adequacy	Avoiding unnecessary restrictions in the other's decision latitude	Sufficient detail and time span of demand forecast; sufficient stability of planning
Monitoring and error diagnosis	Checking the status of execution of the common plan; exchanging information on discrepancies between planned and actual situation	Checking with production that a delivery date can be kept; identifying problem zones in a shared plan
Co-ordination of opportunistic planning	Informing others about implemented deviations from original plan; explicit common decision on modifications	Early information on changes in demands; common decision on plan changes due to an express order
Common reflection / decision for plan cancellation	Recognizing when a plan cannot be executed; reflection on improvement potential	Information on inability to keep delivery date; common reflection on problems in the planning process

Many of the suggested characteristics can in fact be implemented well by direct communication (for instance, communicating anticipated events or negotiating alternatives), but are not limited to direct contacts. Cooperation practices like the exchange of employees between departments can contribute much to the ability of cooperation partners to appreciate the perspective of the other person (*e.g.,* Wehner *et al.* 1996) Also, long experience in a cooperative relationship might lead to a synchronization of actions based on assumptions about what others are likely to do without necessarily requiring direct personal contact (Wittenbaum *et al.* 1996).

5.3 Empirical Evidence for the Model of Collaborative Planning

The model of collaborative planning was explored and tested in two consecutive studies on interdepartmental and supply chain relationships (Windischer 2003; Windischer *et al.* 2009; Günter 2007).

The first study (Windischer 2003; Windischer *et al.* 2009) assessed collaborative planning in three companies characterized by different organizational structures. Collaborative planning processes between the sales, purchase and production departments were compared across companies. The three companies differed in their approaches to handling uncertainties. In company A planning was function-oriented with strong borders between the departments. Sales, purchasing and production departments individually produced and modified plans. Company B was structured in terms of processes in which products or product groups were produced in an integrative manner and where all necessary planning was carried out. In company C the representatives of the departments met periodically each quarter to achieve agreement on the demand forecast. It was expected that there should be differences in the quality of the collaborative planning processes and performance between these three companies given their different organizational conditions.

Purchasers and salespeople in the three companies were asked to describe events in which a person in another company or from another department had done something that was helping or hindering their planning. These critical incidents (Flanagan 1954) were analyzed with respect to the characteristics of collaborative planning described above. Topics in the critical incidents were modification of quantity, delivery date delay, priority modification, critical components procurement, missed deliveries, and technical modifications. As critical incidents might concern several elements of collaborative planning the following percentages add up to more than 100%.

In 40% of the cases in which optimization of the common planning had failed, lack of lateral agreements on goals was identified as a cause, such as the customer not respecting agreed lot sizes or frozen zones. In 38% of the critical incidents, there was a lack of coordination regarding opportunistic planning and communication about revisions. For example, the demand, which had been calculated to cover several months' orders, materialized through orders placed within weeks, and the supplier noticed this change very late. Again 38% of critical incidents involved a lack of adequacy in planning. For example, one partner considered the planning horizon to be too short in relation to the high procurement time for the components. In 23% of the critical incidents, anticipated events were insufficiently communicated. For instance, planning data were too global and did not help provide capacities for future demand. In 17% of the reported cases, the knowledge about the conditions in the reference field was insufficient. Insufficient information on why modifications had happened, lack of exchange about alternatives (*e.g.,* unwillingness to compromise), and lack of monitoring of the common plan (*e.g.,* conflicting orders from the same customer) were also reported.

The characteristics of collaborative planning were combined into an assessment of planning quality. It was found that the quality of the collaborative planning processes was associated with better performance. In line with contingency approaches (*e.g.,* Galbraith 1974; Thompson 1967), low fit between regulation requirements and possibilities seemed to be related to poorer planning and poorer performance. The function-oriented planning in company A did not permit adequate handling of the uncertainties involved in the logistic processes, as indicated by less collaborative planning and poor delivery efficiency. Team-based structures in company B and C on the other hand seemed to support collaborative planning and resulted in better delivery efficiency.

The second study assessed the quality of collaborative planning in two supply chains in forestry (Günter 2007). Survey data were collected from a sample of 107 active supply chain actors, among them forest owners, forest rangers, companies responsible for harvesting and hauling, third party logistics providers, and customers. Descriptive results on collaborative planning revealed communication of anticipated events as posing the greatest difficulties for supply chain actors, whereas mutual agreement on goals and coordination of opportunistic planning seemed to be in line with actors' needs.

In order to learn more about how to facilitate collaborative planning, potential enablers of and barriers to collaborative planning were assessed, specifically interdependence, autonomy, and perspective taking. Results revealed that actors' investments in joint planning depended on the interdependencies given in any supply chain relationship. Actors who felt mutually dependent invested more efforts in collaborative planning than their non-dependent colleagues. Obviously, interdependence gives reasons for actors to engage in collaborative planning because actors feel more committed to the success of the supply chain relationship. Moreover, actors explored their co-actors' perspectives more fully if interdependencies were high. The exploration of co-actors' perspectives helps to bridge differences in cultures and professions, and eases collaborative planning (Parker and Axtell 2001;Vlaar *et al.* 2007).

However, it is important to note that interdependencies do not always stimulate collaborative planning processes. In Günter's study (2007) interdependencies did not affect the level of collaborative planning if actors perceived themselves to be highly autonomous. This individual autonomy might be related, for example, to decision latitude in scheduling delivery due dates and setting performance goals. Contrary to intuitive thinking, autonomous actors invested considerably in collaborative planning irrespective of the perceived degree of interdependence. Even under conditions with presumably few cooperation requirements, that is, high autonomy (based on strategic and/or operative discretion) and low interdependence (based on control of resources and/or the availability of alternative business partners), actors actively promoted collaborative planning. These findings suggest that, in contrast to dominant beliefs in supply chain management, collaborative behaviour can be facilitated not only by tightening but also by loosening interdependencies in supply relationships. This is not to say that loosening interdependencies, per se, is facilitating collaborative behaviour. Langfred (2007), for instance, found that actors may loosen interdependencies in response to conflict, thereby undermining collaborative behaviour. However, if interdependencies are loosened in mutual good faith, that is, in the absence of conflict, this may spur collaborative behaviour, as Günter's findings suggest.

5.4 Views on Planning: A Qualitative Case Study

Planning researchers commonly agree that the more turbulent environments become, the more critical planning is for organizational performance (Mumford *et al.* 2002). However, it is exactly under turbulent conditions, that actors might question planning's usefulness, as revealed in a qualitative case study on the two supply chains in the forestry and timber industry studied by Günter (2007).

Findings from 52 interviews with supply chain actors conducted in this case study indicate that actors' reactions to uncertainties can differ substantially. Among the most common uncertainties in the forestry and timber industry are: natural catastrophes, such as hurricanes, that may uproot the equivalent of several annual yields; bark beetle infestation, that often results in compulsory felling; breakdowns of harvesters and trucks that may cause late deliveries and can increase costs substantially; lack of exchange of planning information, which can undermine coordination efforts; and distrust among supply chain actors, which may lead to opportunistic behaviour in the supply chain and can threaten relationship continuity.

In the face of these uncertainties, some actors, as indicated in the interviews, refrained from planning. Long term planning was argued to be ineffective in the face of an unpredictable environment, which according to interviewees, calls for ad-hoc decision-making. Obviously, for actors who held the view that uncertainties are uncontrollable, proactive behaviours, such as setting up contingency plans, appeared to be useless. However, not all actors shared this view and some reported proactive behaviours and planning efforts. While these latter actors also seemed to be aware of the uncertainties in the industry, they still actively sought to cope with uncertainties, for example, by taking likely disturbances, such as abrupt changes in weather conditions, into account when scheduling deliveries.

Various explanations might hold for these differences in coping strategies. We build upon enactment theory (Weick 1979, 2001) in order to understand these differences better. In a nutshell, enactment theory explains how actors' everyday theories strongly shape their behaviour. This theoretical perspective was already described in more detail in Chapters 2 and 3. Actors who adhere to the "theory" that uncertainties are uncontrollable are not likely to invest in planning. The fewer actors invest in planning though, the less they can see and cut through the uncertainties they experience. Predictability decreases and perceived uncertainty grows. The resulting sense of uncontrollability tends to reconfirm actors' initial preconceptions of an uncontrollable environment which erodes their motivation to plan. The situation may look quite different for actors who think of uncertainties as potentially controllable. They may be more likely to invest in forecasting and contingency planning which might leave them better prepared for the unexpected. If Pasteur is right in his view that "chance favours the prepared mind" (cited from Kaplan and Beinhocker 2003), then it seems plausible to assume that proactive individuals will react more appropriately to uncertainties than their inactive companions.

This view, embedded in enactment theory, adds to the conceptualization of uncertainty in terms of control (see Chapters 2 and 3). The degree to which a situation can be controlled depends on the given predictability and transparency of the situation. If seen through an enactment lens, transparency and predictability appear not to be stable conditions, but malleable and socially enacted. In other words, transparency and predictability may lie, at least partly, in the eye of the beholder. Such different perspectives on the world can be captured graphically by means of cause maps (Weick 1979).

Figure 5.2 shows a cause map for the view on the relationship between planning and uncertainty discussed above. Suppose planning increases. As planning and uncertainty are negatively related, this reduces (perceived) uncertainty which encourages actors to invest further in (collaborative) planning. However, if (perceived) uncertainty increases, actors are more likely to reduce investments in collaborative planning as planning is perceived to be less effective in the face of high uncertainty. The less the actors themselves invest in collaborative planning though, the less controllable the environment appears, that is, perceived uncertainty increases. Actors, who start with the idea in mind that planning might be a useful tool in coping with uncertainties, are more likely to actually gain control. Inactive individuals, by contrast, are more prone to losing control. In other words, proactive individuals focus on planning, not on uncertainties. Inactive individuals, on the other hand, scrutinize uncertainties, not planning's possible virtues. The latter undermines actors' motivation for proactive behaviour and actors are more likely to end up in these seemingly uncontrollable situations that they wished to prevent in the first place.

To recapitulate: actors might question planning's usefulness given high environmental and social uncertainties. However, as has been revealed from this qualitative analysis, actors who refrain from planning can actually bring these uncontrolla ble environments they wish to avoid upon themselves. In order to detect and

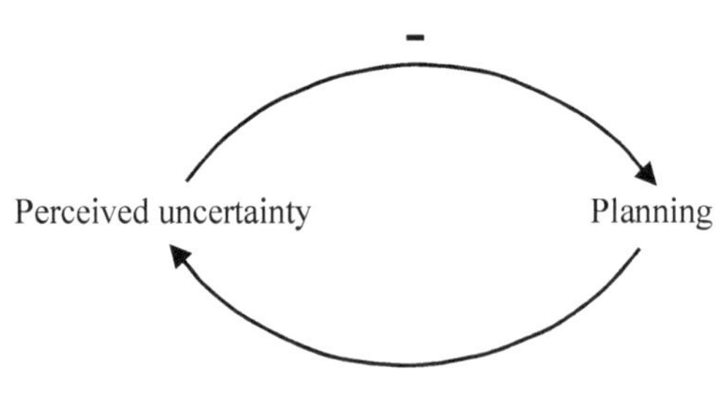

Figure 5.2. Cause map of planning and uncertainty

undo the detrimental enactment processes that may result in uncontrollable environments, practitioners might find it useful to scrutinize preconceptions through making everyday theories visible and using them as a starting point for identifying "unhealthy" assumptions that undermine proactivity and collaborative planning.

5.5 Chapter Summary

In this chapter, planning as a crucial component in managing uncertainties was discussed. Specifically, operational planning within and across organizations was focused upon, which relates to internal logistical processes and external supply relationships. A number of problems in operational planning were identified, especially tensions between central and decentralized planning, autonomy and interdependence, and more and less powerful actors. Some caution was also expressed as regards the general utility of planning and rationalistic beliefs in planning as a blueprint for action. Instead, it was argued that planning should be regarded as a resource for situated action, allowing actors to discover and exploit opportunities for goal attainment and goal adjustment as they arise. In line with much of the logistics literature, collaborative planning was suggested as being central to establishing and maintaining effective relationships within and across organizations. A model of collaborative planning was presented that was developed from the psychological literature on planning. This model contains a number of characteristics which can be used to describe the quality of planning processes, namely communication of anticipated events, knowledge of reference field characteristics, lateral goal agreement, negotiation of alternatives, recognition of planning adequacy, monitoring and diagnosis of errors in the common planning, coordination of opportunistic planning, and common reflection/decision on plan cancellation. In two studies, one on interdepartmental planning and one on planning in supply chains, these characteristics were found to be related to better logistical performance. Also, structures for cross-departmental cooperation, perceived interdependence, and perceived autonomy seemed to foster collaborative planning. Interestingly, autonomy also supported collaborative planning when perceived interdependence was low. Finally, a case study was presented to illustrate the use of cause maps for identifying belief systems as crucial determinants of collaborative planning. It was shown that actors who think of environments as uncontrollable may end up in exactly those uncontrollable situations they wished to avoid.

6

Limits to Control in Automated Systems

“The correct functioning of the train control system and the automatic traffic control system is to be monitored by the signaller. If necessary, he/she has to intervene manually. During normal operation, no monitoring is necessary as long as the operational requirements are met. In the case of disturbances or incidents, the notification of the required services and the required alarm procedures must be guaranteed.” (Excerpt from the rule book of a European railway company.)

The above quote taken from the rule book of a railway company is a perfect example of what has been described as the ironies of automation (Bainbridge 1983). The more automated systems are, the more the human operators have the role of supervisory controllers (Sheridan 1987), implying that they monitor the correct functioning of the technical system and intervene when the technical system fails. At the same time, it may become impossible for the human to adequately fulfil this role due to increasing system complexity in combination with reduced opportunities for practising operational skills and maintaining adequate situation awareness during operation of the system. “[...] the automatic control system has been put in because it can do the job better than the operator, but yet the operator is being asked to (decide) whether it is working effectively. [...] If the computer is being used to make decisions because human judgment and intuitive reasoning are not adequate in this context, then which of the decisions is to be accepted? The human monitor has been given an impossible task” (Bainbridge 1983, p. 776).

The ironies of automation can also be understood in terms of different approaches to managing uncertainty. A strong driver for technological development has always been the increase in control over production processes (Weitz and Shenhav 2000). In this line of thinking, technology is regarded as a source of reliability and safety, as will be discussed in Chapter 7. However, high automation requires a very thorough understanding and high predictability of systems in order to define robust algorithms. This can best be achieved through minimizing uncertainties in the processes to be handled. Whatever tasks are left for human operators to carry out, have to be fulfilled within the tight bounds of technically prescribed operations. As a consequence, human operators are left with very little scope for action and insufficient control over the technical system in the case of disturbances, which may create new risks. Therefore, human-centred approaches to automation (*e.g.*, Hollnagel 2003; Ulich 1998) have stressed the importance of human control even of highly automated systems, especially with regard to the necessity of dealing with disturbances. In line with the coping with uncertainties approach, they argue that human operators need an understanding of the technical processes and adequate opportunities to influence those processes (Parasuraman and Riley 1997). Adequate control furthers motivation and opportunity to develop adequate mental models of the technical system and its processes as well as the operator`s readiness to intervene adequately. More generally, Adler and Borys (1996) have suggested that technology design should follow the notion of an enabling bureaucracy which supports coping with uncertainties.

In this chapter, requirements for human control over automated systems are discussed as prerequisites for holding humans accountable for system functioning. Many methods have been developed to support system design that maintains human control over new technology. The KOMPASS method will be described as one such method. However, possible limits to human control in increasingly complex systems have to be acknowledged. It is suggested that better system design may result when partial non-control is explicitly taken into account. A new approach to design derived from this suggestion is outlined and consequences for handling technological risks more broadly are discussed. Finally, some observations are made on processes of system design and implementation more generally.

6.1 Control and Accountability in Automated Systems

Many authors agree (*e.g.*, Boy 1998; Hauß and Timpe 2002; Hollnagel and Woods 2005; Ulich 1998, Waterson *et al.* 2002) that limits to automation are not only determined by technical feasibility and societal acceptance, but to a large part by the necessity to maintain human responsibility over system goals and their attainment, including all positive and negative (side) effects. Human control over technical systems, including tranparency, predictability, and sufficient means of influencing the systems, is considered to be the main prerequisite for taking on this responsibility. As Hollnagel and Woods (2005) point out, control not only concerns the ability to achieve desired outcomes, but also the ability to recover from disturbances and disruptions. The increasing complexity of automated systems, for instance, through increasing application of "autonomous" and "learning" systems, renders it more and more difficult to meet the conditions for human control, and thereby also raises the question of whether human operators can still be held accountable for the functioning of these systems.

In order to discuss issues of accountability properly, it is important to point out that every automated system is a socio-technical system, independent of its degree of automation, as the workerless factory, the driverless subway, or automated money transfer systems have been developed by humans for humans. Therefore, technical systems should never be looked at in isolation, but always as part of a socio-technical system, which includes humans operating the system and the formal and informal structures and processes within which they work. Furthermore, it is necessary to include in the system definition all those organizations and organizational units that are in charge of system design and maintenance, as well as those that are responsible for rules and regulations controlling system design and operation. This much broader framework helps to reframe questions on the distribution of control and accountability in complex socio-technical systems (Baram 2007).

Before these questions are dealt with further, the KOMPASS method will be presented as an example of current design thinking, which is still focused mostly on supporting humans directly operating with the technology and keeping them in control of that technology.

6.2 Designing for Human Control over Technical Systems: The KOMPASS Method

Many system design methods have been developed whose core tenet is human control over technology (*e.g.*, Boy 1998; Hollnagel 2003; Hollnagel and Bye 2000; Hollnagel and Woods 2005; Timpe *et al.* 2002). The KOMPASS (Complementary analysis and design of production tasks in socio-technical systems) method (Grote *et al.* 2000; Wäfler *et al.* 2003) is presented as a representative example of such methods.

KOMPASS was explicitly developed to further a complementary approach to system design. This approach was suggested several decades ago (Jordan 1963), but has only recently been implemented in design methods (*e.g.*, Boy 1998; Cacciabue 2004; Hoc 2000, 2001; Hollnagel and Woods 2005). The principle of complementarity states that humans and machines are fundamentally different and therefore cannot be compared on a quantitative basis. Instead, they are to be seen as complementing each other in performing a task as a joint cognitive system (Hollnagel and Woods 2005). Functions are not to be allocated to either human or machine. Rather the interaction of humans and machines is to be designed, taking into account reciprocal dependencies between task components and the ways they are carried out by humans and technical systems. Specifically, function allocation should support human strengths and compensate for human weaknesses.

In order to realize a complementary approach in system design it is essential to allocate functions between human and technology in ways that not only allow the human operators to develop and use specific human capabilities, but that also further their motivation to do so. The meaningfulness of a task and thereby also its motivating force is determined by the overall job design. Job design is built on decisions concerning human-technology function allocation as well as division of labour and coordination between different human actors. Therefore, system design must also incorporate the question of human-human function allocation. For instance, the opportunity to programme a system may provide insights into the system's functioning that are crucial for effective supervisory control. Allocating programming and control to different people, possibly even in different organizational units, may jeopardize capabilities for supervisory control. Wall and his colleagues have provided an impressive example of how changes in the allocation of functions between different groups of employees, in this case shifting certain types of maintenance tasks from the maintenance department to workers in production, substantially improved overall system performance (Wall *et al.* 1990, 1992). Hence, a complementary approach to system design must also consider the distribution of tasks among humans as well as among organizational units.

Following the concept of complementary system design, KOMPASS provides empirically tested criteria for an integral consideration of three levels of analysis and design – work system, work tasks of human operator, and human-machine interaction – and support for design teams in pursuing an integral and participative design process. The overall design goal is to enable the resulting socio-technical system to cope with uncertainties, or in socio-technical system design terms to cope with variances at their source (see Fig. 6.1). The notion of loose coupling in terms of balancing minimizing and coping with uncertainties is partially embedded in the design criteria as well. Specifically, the overall criterion for the human task – motivation through task orientation – is a crucial element of loose coupling because it helps to orient the autonomous execution of tasks towards the more general objectives of the organization embedded in the task definition. Furthermore, the criterion of dynamic coupling on the level of the human-machine system aims to balance dependence on the technical system and human autonomy.

6.2.1 KOMPASS Design Criteria

The controllability of the technical system by the human formed the core assumption for the development of the criteria for the *human-machine system*. Grounded in psychological control theories as well as in cybernetics (Sutton and Kahn 1987; Brehmer 1992; Green and Welsh 1988), transparency, predictability and influence were operationalized through four cirteria: process transparency, dynamic coupling, decision authority, and flexibility (see Table 6.1 for some more detail on the criteria).

Criteria related to the *human task* were derived from the concept of motivation through task orientation (Emery 1959; Hackman and Oldham 1976; Ulich 1998). The eight criteria taken from the job design literature for the human task are: task completeness, planning and decision-making requirements, communication requirements, opportunities for learning and personal development, variety, transparency of work flow, influence over working conditions, and temporal flexibility (see Table 6.1).

The *work system* related criteria were derived from the key notion in the socio-technical systems approach that work structure and processes, distribution of tasks and decision authority, and individuals' knowledge and skills should permit the local regulation of system variances and disturbances (*e.g.,* Pasmore 1988). The six criteria for the level of work system are: task completeness (*i.e.*, functional integration), independence of work system, fit between regulation requirements and regulation opportunities, polyvalence of work system members, autonomy of work groups, and boundary regulation by superiors (see Table 6.1).

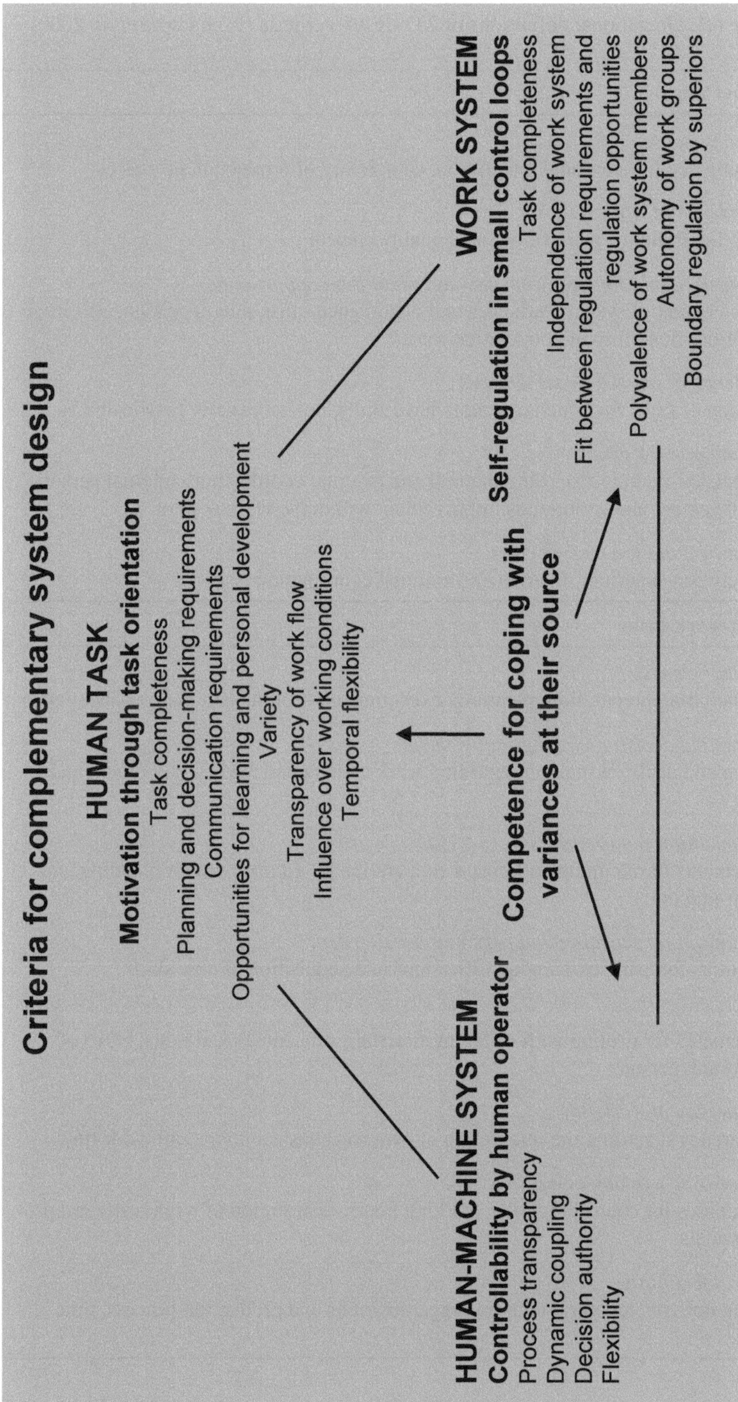

Figure 6.1. KOMPASS design criteria

Table 6.1. Operational definitions of KOMPASS criteria (from Grote *et al.* 2000)

Work system
Task completeness Processing depth, functional integration, complexity of production processes
Independence of work system Buffers, flexibility of work flow, local quality control
Fit between regulation requirements and regulation opportunities Type and extent of system variances and disturbances, task interdependence/form of cooperation, flexibility of cooperation form
Polyvalence of work system members Proportion of tasks for which operators have skills, not necessarily responsibility
Autonomy of work groups Joint decision-making regarding internal and external coordination, internal personnel management and continuous improvement within the work system
Boundary regulation by superiors Ratio between superiors' internal and external coordination tasks
Human work tasks
Task completeness Tasks including preparation, planning, execution, controlling and maintenance/repair
Planning and decision-making requirements Planning and decision-making regarding work content and results, equipment and workflow
Communication requirements Requirements for communication and cooperation based on common planning and decision-making
Opportunities for learning and personal development Possibilities for using existing qualifications and acquisition of new skills
Variety Requirements for dealing with different materials, procedures and tools, types of products and persons
Transparency of work flow Transparency regarding the integration of own work tasks into overall work flow
Influence over working conditions Opportunities for controlling jobs, working hours, distribution of work tasks and production goals
Temporal flexibility Planning horizon, temporal coupling regarding jobs and production process, time pressure

Table 6.1. (Continued)

Human-machine system
Process transparency Opportunities for forming and maintaining mental models of the nature and temporal structure of production processes and of required interventions, process feedback modalities
Dynamic coupling Availability and use of technically provided options regarding time, place, work procedures, and required cognitive effort
Decision authority Distribution of decision authority regarding information access and process control between human operator and technical system
Flexibility Variability of function allocation between human operator and technical system and distribution of the respective decision authority

Differing from other methods of system analysis and design, KOMPASS aims at supporting both the human operators' cognitive readiness to perform their role in the socio-technical system as well as their motivation to use their cognitive capabilities in line with system goals and the responsibility assigned to them for achieving these goals. An example of this is the design criterion "decision authority" on the level of human-machine system. As suggested by Parasuraman and colleagues (2008), authority regarding decisions on information use and decision-making during task execution are distinguished. Specifically, the criterion requires matching information and execution authority such that the human's actual influence over the technical process as defined by the distribution of execution authority makes him or her *want* to use the available information on the technical system as defined by the distribution of information authority. Often automatic systems provide the human operator with a lot of information without assigning him or her functions that would necessitate the use of this information. In incidents that involve lack of acknowledgement of available information, the human operator is often held responsible, thereby neglecting problems of potentially too much information (relating to the discussion of adequate alarm filtering, *e.g.*, Papadopoulos and McDermid 2001; Woods 1995) and of too few incentives to continuously monitor all information available due to lack of action requirements for the human operator.

6.2.2 KOMPASS Design Heuristic

The KOMPASS method also provides a design heuristic which supports interdisciplinary teams in the design of automated systems based on the criteria presented above. KOMPASS covers three phases of design processes: In phase 1 an expert analysis of existing work systems is carried out. In phase 2, an interdisciplinary design team, including system operators, is guided through a conscious reflection of the chosen design approach. In phase 3 concrete design requirements are derived. The sequence, as well as the selection of these phases is not mandatory. For example, if no work system exists that is sufficiently similar to the system being designed, then phase 1 is unnecessary; if a design team is interested in gaining a common understanding of their vision for system design independent of a concrete design project, then phase 2 might be carried out, possibly followed by phase 1 to gain an understanding of how systems have been designed in the past, and so on.

Phase 1: Expert analysis of existing work systems. In this phase, existing systems, which can be considered as a baseline for the design process, are analyzed with respect to the criteria on all three levels of analysis. Guidelines are provided which comprise detailed instructions for data collection by means of workplace observation and semi-structured interviews as well as for data reduction and interpretation with respect to the criteria. Particular attention is also given to variances and disturbances occurring in the system.

Phase 2: Discussion of design philosophy. The main goal of this phase is to establish a common learning process, which enables the team to apply the concept of complementary system design to their concrete problems by reflecting upon their implicit assumptions and theories about the roles of people, technology and organization in work systems. To achieve this, a facilitator leads the design team through six steps, which will be discussed below. Throughout phase 2, the management of uncertainties lies at the heart of the design discussion. By analyzing variances and disturbances in step 2.1, consciousness is raised in the design team regarding the amounts and types of uncertainties in the work system. In the following steps, the underlying question is what role humans and technology should be given in handling these uncertainties.

Step 2.1. Definition of the primary task and the functions of the planned work system. Since the outcomes of a work system always depend on the interplay of people, technology and organization, the aim of this first step is to further a view of the system to be designed as a socio-technical and not just a technical system. To achieve this goal, the design team first defines the primary task of the system as well as the functions of the work system. For each function, potential variances and disturbances are collected in order to identify the functions with high potentials for improvement (step 2.3).

Step 2.2. Definition of a shared evaluation concept to differentiate between successful and unsuccessful work systems. In this step the design team defines a set of design goals with regard to the overall quality of the system. The different expectations of the members of the design team can be made explicit and integrated into a common evaluation concept. This also serves as a basis for discussion and comparison of different design scenarios.

Step 2.3. Identification of the main potentials for improvement. From the collected variances and disturbances, as well as from the system goals defined in the previous step, the design team identifies those system characteristics containing the highest potential for improvement.

Step 2.4. Identification of the potential contributions to a successful work system by human operator, technical system, and organizational conditions. In step 4, the focus is on the principle of complementarity. The design team discusses the different contributions of people, technical systems, and organization for successful system performance. This step aims at rendering evident the complementarity of people, technical systems and organization.

Step 2.5. Specification of the working conditions required for human operators to make their specific contributions. In this step the focus is shifted to the humans' working conditions. Technical as well as organizational prerequisites that enable the human to provide his/her specific contributions to the success of the overall system are discussed. If in step 4, for example, experience has been defined as one of the potential specific contributions of the human operator, the question now is which technical and organizational prerequisites are necessary in order to allow the operator to actually gain that experience.

Step 2.6. Usefulness of the KOMPASS criteria for the analysis, evaluation and design of work systems. In this last step a facilitator introduces the design team to the KOMPASS criteria for complementary system design. In order to initiate a critical reflection on the criteria and the system to be (re-)designed in relation to those criteria, an analysis of an existing system (possibly the one chosen in phase one or another system that is pertinent to the design issues discussed) is carried out together with all members of the design team. Following this common analysis, the results of the expert analysis carried out in phase one will be discussed in the group. The overall goal of this step is to enhance the acceptance of the KOMPASS criteria as support for systematically structuring the design process and to agree on the complementary design approach as the basis for the design process.

Phase 3: Derivation of concrete design requirements. In this phase concrete requirements for system design – and consequently specific demands on the function allocation in the human-machine system – are derived on the basis of the KOMPASS criteria. The KOMPASS method supports the design team in the definition of these requirements, as well as in the definition of concrete work packages to be assigned to the various design experts for their realization.

Step 3.1. Derivation of requirements for system design. On the basis of the KOMPASS criteria, the team formulates design requirements to realize each potential for improvement (as identified in step 2.3). Thus, the problem areas can be systematically analyzed with respect to their potential for complementary design. The goal of this step, however, is not to produce the ultimate technical solutions with all the details set in stone. Instead, the aim is to develop design requirements, which will have to be examined with respect to their technical feasibility and actually realized in a later phase by specialists.

Step 3.2. Definition of work packages. After the workshop, members of the design team are charged with the formulation of design work packages. The aim of this step is to take the interdependent design requirements which were collected in the previous step for each potential for improvement and to combine them to form concrete work packages which can be assigned to the design experts for realization. KOMPASS does not provide specific guidelines for the technical realization of the design requirements. Rather, the realization takes place within the enterprise's usual methodologies for organizational and technical development.

6.2.3 Application of the KOMPASS Method

The KOMPASS method has been applied successfully in a number of system design projects mainly in manufacturing (Grote *et al.* 2000). Following one example is given from the work by Grote and colleagues (2000, p. 280f.) concerning a project carried out together with a manufacturer for welding technology and automation.

The purpose of the project was the redesign of machines for flexible welded mesh production from metal wires. The functionality of these machines includes automated feeding of line wires and flexible resistance welding. The products are mainly used for fences or bent to shape for items of industrial or private everyday use such as shelves or supermarket trolleys. The machines are usually operated by one or two operators who set up the machine, including tool change and adjustment of welding parameters, feed the wire buffers, and supervise the production process. If provided with adequate possibilities, the operators can influence not only the productivity by shortening the production process as well as by reducing set-up times, but also the product quality. The welding parameters need to be adjusted continuously depending on many variables, such as quality of input material, tools (electrodes), machine running time, mesh density, etc. Because these variables are highly interrelated, the process is not (yet) technically mastered. Productivity and quality are therefore highly dependent on the human operators' experience. The focus of the project lay in the design of the interaction of the human operator with the technical system in controlling and monitoring the welding machine.

The analyses of human work tasks and human-machine interaction carried out in *phase 1* revealed weaknesses in several criteria on the level of the operators' tasks.

- Task completeness: The workers' task was not complete because technical planning and job scheduling were not part of their job.
- Planning and decision-making requirements: Simple planning requirements were contained only in the set-up of the machines.
- Variety: The operators' jobs were characterized by little variety with regard to materials, procedures, tools, types of products, and persons they had to deal with.
- Influence over working conditions: The operators did not have any opportunities to influence the scheduling of jobs or the assignment of jobs to people and machines in the work system.
- Temporal flexibility: (a) Planning horizon: Predictability of the jobs was low because the operators were not told in advance on which jobs they would have to work over the next days and weeks. (b) Temporal coupling: The operators could not leave the machines during production because they had to feed them manually.

With respect to the human-machine system the following weaknesses were identified:

- Process transparency: The operators could rely merely on auditory perception and did not receive any feedback from the technical system regarding the quality of the process.
- Dynamic coupling: The operators were tightly coupled to the machines since they had to load them with new work pieces every five to ten minutes. Moreover, monitoring of the process was only possible from one location and the work procedures could not be varied. Finally, the cognitive effort required from the operators could not be influenced.
- Decision authority: Authority over information access and process control was mainly allocated to the machines (*i.e.*, automatic) because the operators had no access to any information source nor were they able to intervene in the process, *e.g.*, by adjusting online the welding parameters. In addition, they were made responsible for the quality of the product but were not supported by the system in influencing relevant process parameters.

In *phase 2*, a two day workshop was organized, during which the project team was led through the following six steps:

Step 2.1: Definition of the primary task and the functions of the planned work system. The design team defined the production of welded meshes as the primary task of the system. The functions necessary to fulfil this task were for example, mesh design, material procurement, job scheduling, machine loading, monitoring of the production process, and quality control. Due to the systemic approach, the design team also considered functions which had never consciously been considered in the design process before. The function "mesh design", for instance, is important for the operator because some of the relevant parameters are defined in that function. It is therefore crucial to include it in reflections on the further development of the welding system. The consideration of the variances and disturbances stressed once again the crucial role of the operator in the system.

Step 2.2: Definition of a shared evaluation concept to differentiate between successful and unsuccessful work systems. The most important criteria to differentiate between successful and unsuccessful systems, as determined by the design team, were: productivity, identification of the operator with the technical system, high product quality, flexibility with respect to different products, system reliability, user friendliness, and support for process innovation.

Step 2.3: Identification of the main potentials for improvement. The design team identified the following potentials for improvement of the system and the production process: mesh definition, definition of the welding parameters, prevention of disturbances, wire feeding and machine preparation.

Step 2.4: Identification of the potential contributions to a successful work system by human operator, technical system and organizational conditions. For each of the design goals identified by the design team in the previous steps, contributions which further or hinder the success of a work system by human operator, technical system and organizational conditions were identified. This step will be illustrated by means of the goal "high product quality":

- Human operator: (a) Furthering contributions: on-line optimization of the process, appropriate education, fantasy, experience; (b) Hindering contributions: sensitivity to distraction, inappropriate job training, de-motivation.

- Technical system: (a) Furthering contributions: reproduction of adjustments, possibility of on-line influencing of the process; (b) Hindering contributions: low transparency, excessive amount of set-ups, lack of mechanical precision.
- Organization: (a) Furthering contributions: procurement of high quality material, appropriate methods for quality control; (b) Hindering contributions: poor opportunities for competence development, lack of quality management.

Step 2.5: Specification of the working conditions required for human operators to make their specific contributions. The design team identified the following pre-requisites for furthering the human contribution:

a) General prerequisites: Operators' opportunities for experiencing success and for acquiring new competences in dealing with the system.
b) Technical prerequisites: Controllability, robustness and transparency of the system.
c) Organizational prerequisites: Decision and action scope within the organization, exchange of experience, praise for goal attainment, opportunities for cooperation within the team, wage system.

Step 2.6: Usefulness of the KOMPASS criteria for the analysis, evaluation and design of work systems. In the reported case, no common analysis with the design team could be carried out. The discussion about criteria took place on the basis of the expert analysis (phase 1). The design team agreed on the usefulness of the KOMPASS criteria as a basis for the derivation of requirements for the system to be designed.

As part of *phase 3*, the following two steps were carried out in the project team:

Step 3.1: Derivation of requirements for system design. The procedure and results of this step will be illustrated by means of one of the identified potentials for improvement: definition of welding parameters. The factors affecting the adjustment of the welding parameters depend on the specifics of the production process. The operator's scope for optimization is quite large. Due to his experience concerning the welding process, he is able to make adequate use of the action opportunities provided. The technical system, therefore, must support the human operator in gaining experience.

With regard to the criteria on the level of human-machine system, the following specific requirements were formulated by the team.

- Process transparency: The operator must be supported by the system when adjusting process parameters, *e.g.,* by displaying default values. Changes in the process parameters performed by operators of the previous shift must be understandable.
- Dynamic coupling: Online adjustment of the welding parameters must be possible and the input of the parameters into the technical system must take place where the operator can directly observe the mesh quality.

- Decision authority: The system supports the operator by proposing possible solutions for the definition of the welding parameters. The operator, however, must have the possibility of accepting or rejecting the solutions proposed by the system.

In a similar manner, requirements were formulated for the criteria on the level of the human task.

Step 3.2: Definition of work packages. The requirements formulated by the design team were grouped into detailed work packages to be assigned to specialists of the different disciplines involved in the design process. On the basis of the results of the workshops, work packages were defined for the following aspects: user interface, definition of the welding parameters, reporting and information system (referring to data about the production process and machine adjustment), possibilities for on-line interaction with the process, wire feeding, welding process, and job scheduling.

These work packages were handed over to the specialists for technical (and organizational) realization on the basis of the company's usual design methodology and project management.

6.2.4 Concluding Comments on the KOMPASS Method

An interesting more recent application of the KOMPASS method happened in the context of a consultancy project, which involved the discussion of scenarios for technological development and their possible effects on the human role in work processes in a railway company (see Chapter 9 for more detail on this project). As part of this project, the design criteria were applied to evaluate task profiles resulting from different options for technology use and accompanying changes in organizational structures and processes. Desirable and undesirable changes in occupational roles could be identified and requirements for technology development formulated. It appeared that KOMPASS was quite useful for broader evaluations of technology scenarios, possibly even more so than for very specific design decisions.

Overall, KOMPASS has proven effective, especially for making explicit and provoking controversial discussions on the fundamental issue of control and responsibility in socio-technical systems. However, it has also become clear that the basic approach of demanding complete human control over technology has to be questioned. This fundamental doubt is expanded on in the following.

6.3 "Human Unreliability": Lack of Human Control over Technical Systems

There is more and more acceptance of the fact that accidents are most often caused by a complex combination of human, technical, organizational, social and environmental factors, which contain latent failure conditions that only through erroneous human action "at the sharp end" will actually lead to negative outcomes (*e.g.*, Reason 1997). Frequently, within this complex interaction of many contributing factors, an unfortunate coupling of human and technology can be found. One example is the accident involving a Lufthansa Airbus A320 in Warsaw on 14 September 1993 (Main Commission Aircraft Accident Investigation Warsaw 1994), where the automatic algorithm (with no manual override) for braking the aircraft after touchdown on the runway was instigated too late. This algorithm is released only when there is a prespecified amount of pressure on both back wheels, which in this case did not happen immediately after touchdown due to a slight tilt of the aircraft and the resulting unequal pressure on the wheels. The delay in braking the aircraft resulted in the aircraft overrunning the runway and crashing into a mound of earth. The aircraft caught fire immediately and two people were killed. As a consequence of this accident, Lufthansa pilots were informed in more detail about the technical definition of the landing procedure and some technical improvements of the Lufthansa Airbus aircraft were implemented. The fully automatic control of reverse thrust and of the brakes during landing was left unchanged. Informing the pilots was meant to re-establish conditions for keeping them responsible, without actually changing the distribution of authority between human and technology.

The problems resulting from removing control from the human operator in this way have been described very well by Amalberti (1992, 1993). He assumes that human operators act on the basis of an "ecological risk management", which allows pilots to deal with their resource limitations by anticipating different courses of action, prioritizing of actions, and exerting active control, forcing the actual situation to follow the anticipated one. He further argues that this way of dealing with risk is rendered more difficult by automation as transparency and flexibility are lost. Pilots react to this difficulty by trying either to outwit the technical system – for example they may enter non-existent wind into the computer in order for the computer to calculate a different, that is, the desired, approach angle – or by fully ceding responsibility to the technical system.

For technology to support the pilots' ecological risk management, system designers would have to consider human situative problem solving strategies more instead of assuming prescriptive optimal strategies. This would require acknowledging the human as being at least as much a safety factor as a risk factor. By viewing the human mainly as a risk factor and assigning the majority of functions to technology as the presumed safety factor, the human is turned into a risk factor instead. A self-fulfilling prophecy is created.

The importance of implicit or explicit assumptions concerning the role of humans and technology in the overall socio-technical system for selecting allocation criteria in system design has been pointed out by Bailey (1989). He distinguished five such assumptions and described their consequences for decisions on function allocation (see Table 6.2). Methods like KOMPASS try to support design teams by systematically reflecting upon their basic assumptions regarding the role of human and technology in bringing about system performance. Specifically, they aim to change designers' assumptions about the nature of humans and technology when necessary, in order to support the human as a resource and safety factor.

Table 6.2. Criteria for human-machine function allocation and underlying assumptions (adapted from Bailey 1989, and Wäfler *et al.* 2003)

Allocation criterion	Implicit assumptions about the nature of	
	Humans	Technology
Cost efficiency The allocation to human or technology is solely determined by who can do the job cheaper.	Both are cost producing factors	
Leftover As many functions as possible are automated, the rest are allocated to the human.	Disturbance and risk factor	Effectiveness and safety factor
Performance comparison Human and technological capabilities are compared and the presumably better performing agent is allocated the respective function.	Both are competing factors	
Humane tasks Functions are allocated so as to create meaningful, motivating tasks for the human.	Valuable resource	Support for human operator
Flexible allocation Functions are allocated redundantly to the joint human-technology system in order to be able to decide on the actual function allocation flexibly according to situational demands.	Both are valuable resources	

6.4 Control and Accountability in Automated Systems Revisited

Beyond assumptions about the nature of humans and technology it is important to also question assumptions about organizations, that is, images of organization (Morgan 1986), especially assumptions concerning possibilities and limitations of planning and control in organizations and the preferred ways of managing uncertainties as has been discussed in detail in Chapter 3. Abandoning the myth of full central control embedded in the minimizing uncertainties approach frees system designers to systematically support decentralized autonomy, constructive coping with the limits of planning, and deliberate choice between local and central control mechanisms. However, for such a change in perspectives on humans, technology, and organization to take effect, another even more fundamental assumption is needed. This assumption states that the most advanced technology would still be controllable by human operators if only system designers and buyers of systems would really want that and were willing to invest more resources in the development of such systems.

After decades of trying to develop theories and methods to effect such a change and as a consequence, design better systems, the recurring discussions on how successful these attempts have really been should be reason to seriously question the assumption itself (Dekker and Hollnagel 2004; Dekker and Woods 2002; Parasuraman *et al.* 2008). What if the insufficient human control of technology is not caused by normative assumptions about humans, technology, and organizations, but by factual limitations of human control and, even more basically, human imagination, due to the ever increasing complexity of technology? Then, either technology development has to be stopped – which is no real option – or the (partial) lack of control has to be accepted. Technically, this acceptance is equivalent to the determination of unmanaged residual risks. On the human side, there is hesitance to admit to a lack of control due to problems of unmanaged accountability. The human operator is kept in the system as a backup where all problems come together and have to be dealt with. The fallibility of this approach and its abuse by system developers and the organizations operating the systems in order not to have to admit to the lack of control has been pointed out by Bainbridge (1983) with utmost clarity. Current concepts of automation in future air traffic management hint at new potential for producing ironies of automation, despite an increasing awareness of human factors issues (e.g., Hoekstra 2001; Straussberger *et al.* 2008; Wickens *et al.* 1998; Federal Aviation Administration 2007; SESAR consortium 2008). The Traffic Alert and Collision Avoidance System (TCAS), which was discussed in Chapter 4 with respect to problems related to incompatible rules, is again used as an illustration because there are also fundamental automation issues involved (see Example 6.1).

Polemically one could argue that the current interest in research on trust in technology at the level of human-machine interaction (*e.g.*, Lee and See 2004; Muir 1994; McGuirl and Sarter 2006; Moray *et al.* 2000; Rajaonah *et al.* 2006) has its roots in the fact that – while still acknowledging that control would be better than trust – trust is all that is left to the human operator. Experiments have shown that technology is trusted most when trust in one's own competences is low (*e.g.*, Lee and Moray 1992, 1994). From general psychology we know that self-confidence is strongly related to perceived personal control (Bandura 2001). Trust may therefore be a consequence of lack of control. This agrees with sociological definitions of trust as a mechanism to cope with uncontrollable uncertainties (*e.g.*, Luhmann 1979, 1988). In the system design literature, however, trust is often understood as a desirable user attitude based on familiarity with the system and confidence stemming from high transparency, predictability, and reliability of systems, thereby actually providing essential prerequisites for control (Lee and See 2004; McGuirl and Sarter 2006; Muir 1994). To explore further the sociological notion of trust as substitute for control, might prove valuable in supporting system design based on the assumption of only partial controllability of technical systems.

Example 6.1. Automation of mid-air collision avoidance

The tragic flight accident over Überlingen on 1 July 2002, has instigated a heated debate on the design of air traffic control systems, in particular the Traffic Alert and Collision Avoidance System (TCAS) (*e.g.,* Weyer 2006). TCAS is a warning system for avoiding mid-air collisions with which all modern aircraft are equipped. When the radar-based system detects a possible conflict with another aircraft, a warning is issued 40 seconds prior to the predicted collision and 15 seconds later the pilots are instructed usually to either climb or descend in order to resolve the conflict. When both aircraft involved in the conflict are equipped with TCAS, the instructions given to the pilots are coordinated and a collision should definitely be avoided. In parallel, aircraft are monitored by Air Traffic Control (ATC), which has raised the issue of assigning priority to either system. With current technology, TCAS information is not directly available to ATC, but has to be transmitted verbally by the pilots. Therefore, it is not unlikely that TCAS and ATC will issue contradictory instructions to the pilots. In the final phase of the Überlingen accident, the collision was caused by the pilots of both aircraft following the rules of their airlines on whom to give priority to. The US crew followed the TCAS instruction to descend, the Russian crew followed the ATC instruction to descend, after having received a TCAS instruction to climb.

After the Überlingen accident, an international standard was released stating that TCAS instructions have priority over ATC and that it is the pilot's responsibility to follow TCAS instructions. Hidden within this standard is the problem that TCAS is not 100% reliable and that, consequently, the pilot has to make the final decision on whether to follow TCAS. Interestingly, the TCAS instructions are also not called instructions as such, but advisories. In the current discussions by aircraft designers about automating TCAS, that is having the TCAS advisory automatically carried out without any action required from the pilot, a manual override is foreseen in order to keep the pilot in charge and also, responsible for executing TCAS instructions. The role of ATC remains unclear. Through automatic data link technology, air traffic controllers will be informed immediately of the TCAS warnings and instructions, but are not to intervene in any way. Collision avoidance, a core ATC task, is therefore delegated more clearly to the pilots and thereby decentralized in these situations, while at the same time it is more automated with less de facto control of the humans involved.

The current situation, as well as the future scenarios for collision avoidance, contains an inadequate distribution of uncertainty and of resources for uncertainty handling. Pilots are to consider TCAS as a completely reliable decision-maker during collision avoidance. At the same time, the availability of a manual override function indicates that pilot input may be required due to TCAS problems and that the responsibility stays with the pilot. The uneven distribution of information between pilots and ATC will be alleviated through data link technology, but ATC is not to use the information for any of its own actions, but leave collision avoidance to pilots and TCAS completely. It remains to be seen whether non-intervention by air traffic controllers in the case of wrong decisions by TCAS and/or pilots will really remain uncontested.

6.5 Designing Partially (Non-)Controllable Systems

If one assumes that technology cannot be controlled fully by human operators, however hard system designers and operators try, the criteria for system design have to be changed. The main purpose of such new design guidelines would be to free human operators of their impossible role of trying to fulfil stop-gap and backup functions in complex socio-technical systems. Methods supporting adaptive system design indicate a move in a similar direction by allocating control fully and without human influence to the technical system in very stressful situations (*e.g.*, Inagaki 2000; Moray *et al.* 2000). However, the crucial issue of assigning responsibility and accountability is usually not dealt with in these design methods.

According to Kornwachs (1999) the main prerequisites for taking on responsibility are the freedom to make choices and to act in a chosen manner. If people are forced to act in a particular way they cannot be held accountable unless they have brought this situation upon themselves. Also, he argues that all necessary knowledge concerning the action, its purpose and its consequences, has to be available and attainable. He emphasizes that automation attempts to reduce complexity for the human operator in order to achieve these preconditions, but that at the same time, new complexities are created which may violate these conditions.

In order to provide the necessary preconditions for taking on responsibility, and thereby also control, the limits of control should be defined as clearly as possible. In those areas which are classified as outside the control of human operators, they cannot be held responsible. Taking the braking procedure in the Airbus A320 as an example, this would mean that the irreversible automation of the essential braking functions should be taught in pilot training and should also be indicated in the cockpit during landing. If mistakes happen in the execution of these functions, the system developer, or possibly the organization operating the system, should be held responsible, but not the pilots. Only if the pilots – in line with Kornwachs' definition – can be proven to have induced this situation deliberately or carelessly (*e.g.*, as a consequence of insufficient competencies) might they have to assume some of the responsibility. Even in such cases, the responsibility might lie more with the airline, especially if a particular pilot's lack of ability or knowledge has already surfaced earlier, for instance, during simulator training, and no action has been taken.

Issues of responsibility and accountability have also become increasingly prevalent due to technologies described as pervasive computing (Satyanarayanan 2001). Pervasive computing refers to tiny devices, possibly embedded in almost any type of object imaginable, including cars, tools, appliances, clothing and other consumer goods, all communicating through increasingly interconnected networks. During a recent workshop on the opportunities and risks of pervasive computing, it was postulated that lack of transparency and inherent uncertainty in these technical systems should be made explicit (Meier 2005). This could imply, for instance, that when switching on an RFID (radio frequency identification) application – such as the milk bottles in our refrigerator that autonomously order new milk when their use-by-date has passed – the user is informed about which network connections have been established and which of those are part of an unprotected network. Such indications do exist already for some internet applications. If the user still has the choice not to use the connection, the responsibility would rest with him or her. In cases where we are forced to use such connections, we would take note of the insecure status and consequently, lack of control, without being able to influence the situation, but also without being held responsible.

In a similar vein, Bellotti and Edwards (2001) have argued that in the design of context-aware systems, issues of intelligibility and accountability must be addressed. Context-aware systems are an important element of pervasive computing as they are able to identify different contexts and change their behaviour accordingly, for instance, allowing or prohibiting connections with other systems. Availability for personal contact in work teams, as an example, might be signalled on the computers or mobile phones of members of these teams based on each of them being in their office and not currently speaking with anybody. Bellotti and Edwards argue that context-aware systems need to be intelligible for users, which necessitates that these systems, before acting upon what they infer about the context must be able to represent to their users what they know, how they know it, and what they are about to do about it. Additionally, accountability must be ensured by providing explicit mechanisms that enforce user identification and action disclosure and by effective user control. The latter does not necessarily imply that the user is intimately involved in task execution but that the desired outcome is attained through an appropriate interaction between system and user.

In order to design socio-technical systems according to such guidelines, existing methods for describing and assessing technical, human, and organizational risks should be extended to clearly indicate zones of uncontrolled risks for both the organizations operating the system and the human operators at the sharp end. Analyzing uncertainties with which the system is confronted and different ways of handling these (reducing, maintaining, or increasing) as outlined in Chapters 2 and 3 should be an important element of such methods. Once zones of uncontrollable risks have been identified, it would have to be decided whether or not functions in such zones could be fully automated and whether human operators can in any way be supported in acting appropriately despite their reduced level of control. Given the unpredictabilities in the defined zones of limited or no control, support for operators could only be in the form of heuristics. Process rules as suggested by Hale and Swuste (1998) could be such heuristics, that is, rules which do not specify a concrete action nor only specify the goal to be achieved, but provide guidance on what to do in order to find out what the right action is in a given situation. This could concern, for instance, information sources to be used, other actors to be involved, or definition of priorities (see also Chapter 4).

At the same time, in these zones of limited control the responsibility for the safe operation of the system would remain with the system developer and potentially the organization operating the system, but not the human operator. The pressure to keep these zones small and thereby maximize control for the human operator would increase. Something similar is achieved already by US law allowing system operators to sue the system developer when his or her own erroneous action can be proven to be a consequence of bad system design (Baram 2007).

A method for risk analysis that could be helpful for identifying zones of no or limited control is Hollnagel's (1998) CREAM (Cognitive Reliability and Error Analysis Method), which has recently been developed further into an Extended Control Model (Hollnagel 2007). In Hollnagel's model four modes of control are distinguished: scrambled (choice of next action is close to random), opportunistic (choice of next action driven by narrow focus on current situation), tactical (choice of action based on following a known procedure) and strategic (choice of action determined by higher-level goals with longer time horizon). Control may concern anticipatory or compensatory action with each involving more specific activities of targeting, monitoring, regulating, and tracking. The four modes of control are assigned human error probabilities which are used to determine the overall probabilities of human error in event sequences potentially involving several switches between the different modes. By analyzing possible event sequences for a particular human-machine system and the different control modes involved, zones of no or scrambled control can be identified and appropriate measures taken to handle processes in these zones. As an example of such an approach which already exists, one might take the 30 minute rule in nuclear power plants. This rule demands that after the occurrence of major non-routine events the process control systems in these plants are capable of keeping up a sufficiently safe level of operation for as long as 30 minutes without human intervention, thereby giving the human operators time to recover from a state of confusion and scrambled control and to regain tactical or strategic control. Also, different stages of action regulation have to be distinguished, that is, information seeking, decision-making, execution of decided actions, feedback on effects of action, and corrective actions. The lack of control may only concern some of these stages and also, may affect different stages differently. The outcome of a CREAM based analysis would help to make decisions on full or partial automation more systematically, aiming at a very deliberate match between control and responsibility. Also, instead of pretending that systems are safer than they are, in particular due to their increasing embeddedness in complex networks, it may be much better to regain overall control by admitting to areas of opaqueness and uncertainties in the system. Only then can the ability to cope with uncertainty and loss of control be trained and supported systematically.

6.6 Towards a New Politics of Uncertainty in System Design

Much of what has been said in this chapter is not new, but the underlying attitude may be: Instead of lamenting the lack of human control over technology and of demanding over and over again that control be reinstated, the approach presented here states very explicitly that current and future technology, even with the best of system design, contains more or less substantial zones of no control. Any system design should build on this supposition and develop concepts for handling the lack of control in a way that does not delegate the responsibility to the human operator, but holds system developers, the organizations operating the systems, and societal actors accountable. This could happen much more effectively if uncertainties were made transparent and the human operator was relieved of his or her stop-gap and backup function. This would also bring us closer to Kornwachs' (1999) requirement: act in a way that keeps the conditions for responsible actions of all involved intact. In order for such a change in perspective on system design to be successful, fundamental changes in regulatory policy and public attitudes have also to be effected. In line with Power's call for establishing a new politics of uncertainty (Power 2004), we all have to learn to accept that risks can, at best, be responsibly handled, but that they cannot be managed away.

In his keynote address at the Academy of Management Conference 2007 in Philadelphia, Charles Perrow expressed his regret about people having taken inadequate handling of risks to be the main message of his book *Normal Accidents*. He stated that, in fact, his main concern had been, and still is, the inappropriate use of power when making decisions about risks. Decision-makers in companies and in governments to a large part determine the levels and distributions of risks that we as a society and as individuals in various roles are confronted with. Of course, within democratic societies there is, at least, some participation from other stakeholders in these decisions, but the actual power for making decisions on risks is distributed very unevenly.

One crucial element for handling this power differential appropriately is transparency in the decision-making processes related to developing, implementing, and using automated systems. There should be transparency of the decision criteria used, the methods applied to determine measurements on these criteria, and the consequences of different decision alternatives for all stakeholders. Given the highly conflicting interests involved in decisions especially regarding technology in high-risk systems, power is often employed to smooth over or even cover up conflicts in the interest of the most powerful stake-holders. As a consequence, governments and companies are strongly distrusted by the public, which in turn decreases the level of openness in risk-related communication even further. But even if sincere efforts are made to be transparent, there is the additional problem of knowledge differentials regarding the highly complex issues involved in such decisions. These knowledge differentials are often used as a – partly defensible – argument for not providing all possible information during a decision-making process to all the stakeholders involved.

Another equally important issue is the accountability for decisions taken. Organizations and governmental agencies having the power and presumably also the knowledge to make sensible decisions on risk involved in using particular technologies creates the rightful assumption that they can also be held accountable for the decisions they take. In recent times, the public has demonstrated that it can exercise significant power also – especially in their role as consumers – when organizations do not live up to their responsibility. Several organizations have been damaged severely as a consequence of their inadequate and self-serving decision-making. Shell and their decision to dispose of the Brent Spa oil platform in the North Sea is probably still the most prominent example. Power (2004) describes the ensuing concern with reputation management as another severe problem, because risk experts now frame their judgments more in terms of reducing their personal, legal and reputational risks than in terms of providing honest assessments of the risks at hand.

The postulated new politics of uncertainty "would not seek to assuage public anxiety and concerns with images and rhetorics of manageability and control, and would challenge assumptions that all risk is manageable. States and corporations would not need to act as if all risk is controllable and would contest media assumptions to that effect. Public understandings of expert fallibility would be a basis for trust in them, rather than its opposite" (Power 2004, p. 63). By taking on this new approach to uncertainty on a societal level, it is hoped that decisions on the design of automated systems can be improved in the future. Instead of taking the minimizing uncertainty approach for granted uncertainties and risks involved in introducing technology need to be openly discussed. Design decisions have to be based on explicit scenarios which demonstrate how adequate coping with these uncertainties and risks can be achieved, including the acknowledgement of limits of controllability and the definition of responsibility for the functioning of automated systems within and outside these limits.

6.7 A Note on the Process of System Design and Implementation

When thinking about how systems should be designed, it is also important to reflect upon the general role that technology can and should play in organizations. In order to so, the basic requirement is to decouple technical systems from the tasks they are to perform when defining technology. In earlier sociological discussions, technology was defined as the work people do (*e.g.,* Comstock and Scott 1977; Perrow 1967), which impedes an independent consideration of the effects different designs of technology can have on people`s tasks and thereby also renders it difficult to derive design requirements, as acknowledged explicitly by Perrow himself a few years later (Perrow 1983).

The design principles discussed in this chapter are grounded in the assumption that technology is always an independent as well as a dependent variable in system design. That is, technology changes organizations, but it is also deliberately chosen and changed by organizations. Neither technology nor organization can be considered the dominant causal agent (Markus and Robey 1988), but their reciprocal interaction will bring about change. An early review of the research in this field conducted by Robey (1977) is particularly interesting in the current context of uncertainty management. He argued from the existing evidence that technology acts as a moderator in the relationship between environmental uncertainty and organizational structure. Computing technology supports decentralized structures in organizations that face many uncertainties, while it supports centralized authority in organizations with rather certain environments.

An interesting more recent example for the reciprocal interaction between organization and technology is communication technology. Some authors have argued that this technology will "automatically" further decentralized and democratic decision-making due to its capability to support lateral coordination within and across organizations and even societies (*e.g.,* Sproull and Kiesler 1991). As Clegg and colleagues (2006, p. 384) have phrased it: "the pen and the typewriter gave us bureaucracy; can virtuality give us democracy?" The empirical evidence indicates that electronic communication systems can in fact have democratizing effects given a number of technical as well as organizational prerequisites such as management support for less hierarchical decision processes and task-induced communication and cooperation needs (*e.g.*, Grote and Baitsch 1991). The latter may even further the use of electronic communication for more self-regulated cooperation when management support is not present. Lasting changes in work processes and structures require a strong commitment to increasing self-regulation by central members of management, however. This is a case in point of how organizations strongly influence whether and how a technology's potentials are realized or suppressed.

Generally, the interaction between organization and technology has been studied most intensively from the perspective of adaptive structuration, which stresses human agency in selecting, interpreting and enacting opportunities created by technology (Boudreau and Robey 2005; DeSanctis and Poole 1994; Markus and Robey 1988; Orlikowski 1992, 2000). Usually focusing on information technology, the major assumption underlying this perspective is that the uses and consequences of technology emerge mostly unpredictably from complex social interactions. Therefore, understanding the use of technology and its effects on a particular organization requires intensive study of the actual practices and routines evolving (DeSanctis and Poole 1994; Orlikowski 2000). From this also follows the fact that effects of the same technology will vary in different settings. In his classic study, Barley (1986) examined the introduction of computerized tomography in two radiology departments and the ensuing changes in patterns of interaction between radiologists and technical staff. He found that quite different changes occurred because the CT scanners – instead of determining change by their inherent characteristics – "[...] became social objects whose meaning was defined by the context of their use" (Barley 1986, p. 106). The new technology affected skills and competencies required and consequently, behaviour patterns of seeking and giving advice between radiologists and technical staff. Eventually, these patterns became institutionalized as formal decision-making structures centring on the radiologists in one case and the technical staff in the other.

Technology from this viewpoint presents opportunities for structural change that can be expressed as the spirit of the technology (DeSanctis and Poole 1994). This spirit is related to the designers' intentions, but is not fully determined by them. Rather, it is a combination of the design metaphor underlying the system (*e.g.*, electronic chalkboard), the specific features the system possesses and how these are named and presented, the appearance of the system through the chosen user interface, and the procedures for system instruction and training. The technology's spirit will influence, but not determine, which of many possible alternatives will actually emerge from the process of structuring. As Boudreau and Robey (2005) have put it, the technology gets reinvented through its use, which in their case concerned an enterprise resource planning (ERP) system and therefore a technology that is usually considered as particularly constraining for human agency.

Contrasting this emergent perspective with the perspectives of technological determinism versus organizational imperative provides interesting insights into the question of accountability discussed earlier (Markus and Robey 1988). When technology is considered as a causal agent producing predictable impacts in organizations, then responsibility for these effects lies with whoever made the decision on which technology to introduce where. Which changes are created can only be influenced by stopping, slowing or accelerating the introduction of a particular technology. From an organizational imperative position, managers and designers have a strong influence on the effects of technology, for instance, by making decisions about the specific features of technical systems, improving the design of features, and different implementation strategies and processes. Consequently, all people involved in making and carrying out these decisions are accountable for the technology's effects in the organization. The emergent perspective is less clear on the distribution of responsibility because the effects of technology are attributed to an unpredictable interaction of technological features and actors' intentions and behaviours. If effects are truly unpredictable then systematic intervention is not possible and responsibility for the technology's effects cannot be assigned. Recently, this issue has been taken up by Suchman (2002) in her discussion of located accountability. She argues that the fact that knowledge and practice are distributed and that no one has the overall picture does not relieve anyone of responsibility, but makes everybody personally responsible for what lies in their reach. In line with this view, Bellotti and Edwards (2001) have demanded explicit mechanisms that enforce user identity and action disclosure in order to further personal accountability in context-aware systems, as was mentioned earlier.

Furthermore, it should also be pointed out that the proponents of the emergent perspective seem to assume that user participation in technological change will make the emergent processes of adaptive structuration more predictable, which in turn results in joint responsibility for the technology-in-practice. As Orlikowski (2000, p. 423) states "when the institutional conditions are nonhierarchical and participative [...] the enactment of the collaboration technology-in-practice helped to improve the shared and collaborative design efforts of the developers". This quote does appear to be cautiously admitting that more systematic technology development and use is possible through user participation as has also been demonstrated empirically (*e.g.,* Lin and Shao 2000; McKeen *et al.* 1994). A concrete example for particularly effective participation practices is the study by Edmondson and colleagues (2001) mentioned earlier. Successful change of routines through new technology – in line with the expected use of the technology – was furthered by involvement, which furthered the teams' motivation, by preparatory practice sessions and early trials, both helping to create psychological safety while encouraging new behaviours, and by reflection, which promoted shared meaning and improvement initiatives.

Irrespective of whether one wants to follow the perspective of adaptive structuration and technology-in-practice in system design, or, on the other hand, has a more deterministic view of the reciprocal interaction between technology and organization, one will be confronted with the necessity of gaining a thorough understanding of the technological potentials as well as the forces operating in the organization in favour of and against realizing these potentials (Klein and Sorra 1996). These forces are very strongly linked to assumptions about appropriate management of uncertainty and about the role humans can and should play in work systems including their interaction with technology in fulfilling this role. Methods such as KOMPASS aid reflection on these assumptions and how they affect the design and implementation process. Also, there should always be a high degree of user participation in order to bring in different perspectives on technological options and technology-induced organizational change. This is not to say that user participation automatically improves system design. In fact, obtaining reliable information on user requirements and integrating this information into a system definition is far from easy and will mean increasing uncertainties instead of reducing them during the design process. These difficulties have been much discussed in system design (*e.g.,* Shaw and Gaines 1996) as well as more generally regarding knowledge elicitation and sharing (*e.g.,* von Krogh 1998). However, the effort is worthwhile as well-managed user participation will result in better and more accepted systems (*e.g.,* Lin and Shao 2000; McKeen *et al.* 1994).

Finally, this last point illustrates that the design process itself can be designed according to different approaches to managing uncertainty. How much participation by different stakeholders and how much openness for different solutions is sought, for instance, are indications of different ways of handling uncertainties. A design process can be turned into a mechanistic fulfilment of project management requirements aimed at minimizing project risks. It can also be an open search for novel solutions acknowledging the necessity to handle potentially vast amounts of uncertainties in order to achieve innovation.

6.8 Chapter Summary

In this chapter, design of automated systems was discussed by elaborating on the issue of control and accountability in such systems. Regardless of a system's degree of automation, it is humans who are responsible for its functioning. The basic tenet of human-centred automation – that human operators need to be in control of technical systems – is derived from this responsibility, referring back also to the necessity of enabling uncertainties to be coped with in order to balance stability and flexibility in highly automated systems. KOMPASS was presented as an example of methods aimed at supporting managers and engineers in designing systems that are controllable by the human operator, especially in those conditions where all other safeguards fail. In view of human fallibility in general, as well as evidence of loss of control over technical systems stemming from incident and accident analyses, the question was raised how achievable the goal of human control over technology really is. Reasons for lack of controllability can be found in the normative assumptions of those developing and implementing technology which may support a self-fulfilling prophecy of turning humans into risk factors. However, the ever increasing complexity of systems also has to be acknowledged as limiting control. It was therefore suggested that system design be founded on the premise of partial non-controllability of technology. This approach could help human operators to deal better with system opaqueness and uncertainty by providing systematic information on the limits of control and thereby also relieving them of some of their responsibility. At the same time, this approach would force system designers, the organizations operating the systems and regulatory institutions to take on responsibility for use of technical systems whose complexity can no longer be mastered entirely. Consequences for making decisions on system automation by regulators and companies were discussed within the larger realm of establishing a new politics of uncertainty on a societal level. Finally, some comments were made regarding the basic relationship between technology and organization and the limitations on systematically establishing particular use patterns through technology design.

7

When Uncertainty Implies Safety Risk

The most common understanding of safety is the absence of bad things happening, be it accidents, illness, pollution or global warming. As absolute safety in terms of complete absence of bad events is impossible, safety is usually framed as acceptable risk, where risk is defined as the product of probability and damage. The combination of safety and risk into the term safety risk is often used to denote risks which are not financial risks or political risks, but risks to human health and the environment.

In the following, organizations' attempts to handle safety risks by means of safety management are discussed within the framework of uncertainty management. As the uncertainty in this case always concerns events to-be-avoided, there is a particularly strong tendency to follow the minimizing uncertainties approach. However, there is an increasing awareness that coping with uncertainty is also needed. This need for flexible adaptation has been discussed, especially by proponents of the High Reliability Organizations (HRO) approach (*e.g.,* Roberts 1993; Weick *et al.* 1999) and more recently by system engineers in terms of resilience engineering (Hollnagel *et al.* 2006, 2008).

An important distinction to make when discussing safety is that between process safety and personal or occupational safety (Grote 2007b). In process safety, the risks and uncertainties to be managed are directly linked to the primary work process of the organization, such as curing patients, producing energy or transporting goods and people. Potential damages result from failures in the execution of tasks linked to this process, without necessarily harming the human operators involved. The death of a patient due to false medication would be an example here. In personal or occupational safety, potential damages concern the human operators, but in turn are not necessarily directly linked to the primary work process. The most frequent occupational accidents, such as falls, cuts, and bruises may result from falling down stairs on the way to work or getting stuck in a door, just as much as from being hit by a heavy tool or getting caught in a machine. Personal protection against these hazards is mostly a secondary task, sometimes even interfering with the primary task, as, for instance, in the case of ear protection preventing detection of task-relevant signals such as unusual machine noises. Accordingly, measures to promote process safety versus personal safety may have to be different, just as much as indicators of successful interventions need to be different. A decrease in cuts and bruises does not say much about the process safety standard. On the other hand, there may be few incidents concerning process safety and still a high number of occupational accidents.

Often the safety management in organizations does not distinguish sufficiently between these two types of safety. Also, in the academic literature there has only very recently been a push for making this distinction more prominent as a consequence of the BP Texas City explosion (Hopkins 2009). In this particular case, the investigators found that BP management had emphasized occupational safety and rewarded supervisors for low work accident rates in their teams, which led to a very good occupational safety record. At the same time, many glaring warning signals regarding declining process safety were not reacted to or even taken note of. To date, there is insufficient conceptual and empirical knowledge on how measures aimed at process safety and occupational safety interact. Therefore, as a first step, it is important to clarify which type of safety is the focus for particular measures, how they are assumed to affect safety and how their effects are evaluated.

In this chapter, the focus will be on process safety which is more directly linked to the design of work processes and to managing uncertainties embedded in these processes. After a brief example illustrating the relationship between process safety and uncertainty management, this relationship is discussed in more depth regarding advantages and disadvantages of following a minimizing versus coping with uncertainties approach. As in the previous chapters, a strong case is made for achieving a balance between the two approaches through loose coupling. The concept of safety culture is introduced in terms of its potential for supporting this balance. Because much of the reluctance by practitioners and regulators alike against increasing capabilities for coping with uncertainty is related to assumptions about the relationship between autonomy and safety, research on this relationship is reviewed in detail. Finally, a procedure is proposed for the assessment and design of safety management and safety culture from an uncertainty management perspective, drawing on the decision process suggested in Chapter 3.

7.1 A Case of Handling Uncertainties: Process Upset in a Polyethylene Plant

In the plant of a large petrochemical company, polyethylene is produced from ethylene dissolved in isobutane and a number of other chemicals including hexane. The reaction takes place in loop reactors under high temperature and pressure. An operator in the plant's control room monitors two such reactors by means of a number of screens and process recorders on a control panel. Looking at one of the process recorders, another shift operator explains to the observer that when two of the curves on the line recorder do not run in parallel anymore, extra caution is needed, and when the curves cross, the process has to be stopped immediately. Those two curves concern the pressure in the reactor and the energy consumption in a group of pumps. The crossing of the curves indicates lumping of the polyethylene in the reactor, which increases the pressure in the reactor and the energy consumption by the pumps because more energy is needed to pump the finished product out of the reactor. Next to the process recorder, a piece of paper is taped to the control panel, stating critical values for these two parameters, distinguishing between values when the shift supervisor has to be informed and when the process has to be stopped. Stopping the process would entail the immediate emptying and rinsing with water of the reactor and an interruption of production for several hours.

An hour later, during the course of the observation, the curves do indeed begin to move towards each other. The panel operator notices the change immediately and changes the set values for hexane after having checked a number of other process parameters and having also verified the set values for hexane in the standard operating procedures. This action causes the process control system to reduce the influx of hexane which reduces the pressure in the reactor due to a smaller volume of reacting substances. At the same time, the operator has informed the shift supervisor who leaves a meeting to join him at the control panel where he remains during the course of the process upset. The first actions taken by the operator have not been able to reverse the trend in the two parameters. Only after further reduction of hexane influx and more rapid emptying of the reactor do the values return to normal.

In the fifteen minutes that this course of events takes, the curves displayed on the process recorder have briefly crossed. Trusting his own competence in handling the process upset and supported by the shift supervisor, the operator decides against stopping the process completely. Instead of causing a significant interruption of production, the operator succeeds in normalizing the process in the course of half an hour. The results from quality control are positive again a little while later also. His shift colleague comments: "I definitely would have stopped the process completely", but admiration for the other's competence can be sensed.

When safety professionals or managers in general discuss this example – as they have in many of the courses I have taught – there are usually mixed feelings. Some participants argue that important information, for instance, on the technical defences installed, is missing and therefore assessment is impossible. This is the easy way out and does not take into consideration that people in real situations often also have to make decisions based on incomplete information. Unfortunately, it is not all that uncommon that process operators in such plants do not have sufficient knowledge about technical safety margins. When people are willing to comment on the case, there are usually some who are strong believers in minimizing uncertainty using rules which would sanction the operator and possibly also the shift supervisor for violation. There are always some, though, who acknowledge the competent handling of the upset and are more worried about unclear signals regarding the support for coping with uncertainty as indicated by the very different attitudes expressed by the two operators. Both positions will be discussed further in the following sections, again in terms of the question of finding the right balance between the two approaches to uncertainty.

7.2 Minimizing Uncertainties as the One Road to Safety?

The minimizing uncertainties approach is especially prevalent in organizations where malfunctioning not only endangers business success but also human life and environmental integrity because the production processes incorporate major risks for health, safety, and environment. Usually in these organizations not only is the chosen management style that of minimizing uncertainties, but this approach is even prescribed by external regulatory bodies as part of zero-risk regulation (Amalberti 2001; Gilbert *et al.* 2007; Kirwan *et al.* 2002). At the core of the minimizing uncertainty approach to safety lies the assumption that humans are risk factors who – if it is not possible to make them redundant by automation – have to be restricted as much as possible in their freedom to act in order to channel their actions towards the one safe way of doing things.

But even for these systems, it has been argued that the minimizing uncertainty approach is less than optimal due to the system's reduced capability to adequately act in the face of requirements stemming from internal and external disturbances (Perrow 1984; Weick *et al.* 1999; Hollnagel *et al.* 2006). An interesting result which indirectly supports this claim is that the effectiveness of safety training in reducing accidents and injuries and increasing safe work behaviour appears to be influenced by the general tendency to avoid uncertainty (Burke *et al.* 2008). Safety trainings undertaken in countries that, according to Hofstede's (2001) classification of national cultures, are considered to be highly avoidant of uncertainty were less effective, which the authors explained in terms of the safety trainings themselves being more structured and concerned with avoiding uncertainties, thereby not providing sufficient opportunities for trainees to learn how to handle non-routine situations.

Other examples stem from studies in healthcare. Klemola and Norros (1997), for instance, pointed out the importance of recognizing uncertainty as a basis for effective practice in anaesthesia. They argued for promoting what they called a realistic orientation that relies on situated information search and action in contrast to an objectivistic orientation implying a deterministic implementation of a preoperative plan. Similarly, Faraj and Xiao (2006) described two modes of coordination in shock trauma teams, expertise coordination and dialogic coordination. While the first mode relies on preset plans and protocols, the latter mode involves sensemaking activities and potentially even protocol-breaking in order to respond adequately to non-routine events.

The question remains, though, and has been frequently restated (*e.g.*, Dekker 2003; Rasmussen 1997; Woods and Shattuck 2000), of how to improve the predictability and controllability of high-risk systems while at the same time increasing their flexibility in the face of tightly coupled subsystems that may interact in unforeseen ways. Perrow (1984) has argued that this represents a dilemma that cannot be solved and consequently some systems cannot be run safely, which is core to his normal accident theory. Others have been more optimistic about overcoming this dilemma, notably researchers advocating the High Reliability Organization (HRO) approach (*e.g.*, Roberts 1993) or the more recent concept of resilience engineering (Hollnagel *et al.* 2006).

Also, in regulatory approaches, the balance between minimizing and coping with uncertainties is under debate. There seems to be a general move away from prescriptive regulation specifying in great detail how an organization has to carry out its operations towards goal-oriented legislation, which only prescribes the goal to be reached and leaves open the way in which this is accomplished (Kirwan *et al.* 2002). Goal-oriented legislation can be seen as furthering self-regulation within organizations more in line with the coping with uncertainties approach (Majumdar and Marcus 2001). This also links back to the discussion in Chapter 4 about different rule types. Goal-oriented legislation would imply more goal rules as compared to process or action rules (Hale and Swuste 1998).

On an even higher societal level, there has been a prevalence of so-called zero risk regulation, that is, attempting to or at least making the public believe that attempts are made to reduce all risks down to (almost) nothing to worry about. Thus, we all happily believe that we can benefit from various high-risk technologies and practices and still be safe. Recurring disasters in all types of high-risk operations have awoken us from that dream numerous times, which has led to a discussion on the feasibility and appropriateness of zero risk regulation and the possibility of more openly admitting to risks in order to find more appropriate ways of handling them (Gilbert *et al.* 2007; Power 2004). This suggested shift in safety and risk policies would again permit a move away from a purely minimizing uncertainties approach, as was also discussed in Chapter 6 regarding the design of automated systems.

7.3 Safely Coping with Uncertainties?

For the last twenty years, an eminent group of researchers has studied a multitude of high-risk systems posing the generic question of how these systems manage to run their operations safely despite the high risks involved and despite the organizational dilemma of being stable and flexible at the same time (*e.g.*, Roberts 1993). They called these organizations High Reliability Organizations (HRO) and identified a number of characteristics that seem to make these organizations special (Weick and Sutcliffe 2001):

- preoccupation with failure, *e.g.*, through intensive analyses of accidents and incidents;
- reluctance to simplify interpretations, *e.g.*, by encouraging diversity of opinions;
- sensitivity to operations, *e.g.*, by monitoring normal operations for signs of imperfection in the organization's defenses;
- commitment to resilience, *e.g.*, by supporting recovery from errors through improvisation based on the acknowledgment that errors can always happen;
- deference to expertise, *e.g.*, by giving decision authority to the most competent person irrespective of their formal rank.

These characteristics are assumed to further a particularly high level of mindfulness, which is needed to build the capability to discover and manage unexpected events, which in turn leads to higher reliability (Weick *et al.* 1999).

From the many case studies conducted on HROs, three results are particularly noteworthy:

1. HROs are able to *switch between more centralized and more decentralized operation*. This was one of the core findings from the initial HRO studies on aircraft carriers (LaPorte and Consolini 1991). Three modes of operation were found: centralized, hierarchy-based operation under normal conditions, decentralized competence-based operation under conditions with increasing uncertainties, and fully pre-specified emergency operation.
2. HROs have systems in place that *very deliberately manage the balance between stability and flexibility*. In their study, previously mentioned, on the incident command system in a fire department, for instance, Bigley and Roberts (2001) found specific processes that helped to rapidly alter formal organizational structure in response to changing demands. There was structure elaborating (the actual forming of the emergency response organization on the spot), role switching (reassigning personnel to different positions in line with changing task demands), authority migrating (decision-making authority given to whoever was the most competent for a particular decision), and system resetting (disengage the system when the chosen strategy was not effective or the situation changed dramatically and reconfiguring it on the spot).
3. HROs are characterized by *strong cultures that serve as a centralization mechanism* in an otherwise decentralized organization (Weick 1987). This particular point will be taken up in more detail in the section on safety culture.

More recently, the concept of resilience engineering as a less descriptive and more design oriented approach to balancing the needs for stability and flexibility in response to uncertainty has emerged (Hollnagel *et al.* 2006). The term resilience has its origins in biology and ecology, denoting in most general terms the ability of systems to recover from perturbations (Brand and Jax 2007). One earlier use of the concept in the safety domain can be found in Wildavsky`s 1988 book *Searching for Safety*, where he discussed different strategies for handling risks on a societal level by contrasting anticipation and resilience. While anticipation attempts perfect prediction and is the basis for much legal regulation, such as "no trials without prior guarantees against error" (Wildavsky 1988, p. 9), resilience relies more on fast adaptation to emerging threats and decentralized small-scale trial and error. He argues that sometimes a mixed strategy may be useful in the form of decentralized anticipation, that is "numerous independent probes of an uncertain future [...] (which provides) the ability of market competition to interrogate the unknown at a low cost to society, while simultaneously encouraging individuals to overcome adversity" (Wildavsky 1988, p. 8). In a similar vein, Power (2004) has argued for the establishment of new politics of uncertainty, as discussed in Chapter 6, which as one crucial element "must generate legitimacy for the possibility of failure". He assumes that public trust could be reestablished "precisely because an explicit disclosure of possible failure is embedded in innovatory processes" (p. 62).

Resilience engineering applies the concept of resilience to the design of organizations as socio-technical systems (Hollnagel *et al.* 2006, 2008). Woods has defined a number of indicators of resilience: buffering capacity, flexibility as adaptive capacity, sufficient margins, and tolerant behaviour near system boundaries (Woods 2006). From these indicators, design requirements can be derived. Given that resilience is about buffers and slack, thereby reducing system efficiency, an important issue for system design is which kinds of "inefficiencies" increase resilience, and which are simply inefficiencies. However, one can question whether resilience is actually achieved by design or rather by management activities (Wreathall 2006) in order to further a basic attitude towards dealing with system failures in line with the commitment to resilience postulated by the HRO-researchers (Weick *et al.* 1999). Such safety-related attitudes have been much discussed in recent years under the heading of safety culture, which is the focus of the next section.

7.4 Safety Culture as a Mechanism for Loose Coupling

Organizational culture has been defined as "a pattern of basic assumptions – invented, discovered, or developed by a given group as it learns to cope with its problems of external adaptation and internal integration – that has worked well enough to be considered valid and therefore, to be taught to new members as the correct way to perceive, think, and feel in relation to those problems" (Schein 1985). In the context of the the debate on balancing stability and flexibility in high-risk organizations, Weick (1987) has pointed out that culture serves as a strong basis for a form of coordination and integration that incorporates both decentralization of autonomy and centralization of values and norms as binding forces for local action. Through this conception of culture, the often rather superficial discussion of safety culture as safety-promoting norms and attitudes shared by the members of an organization appears in a new light. Culture from this perspective is seen more as a means of providing sufficient coordination and integration of otherwise autonomous agents – thus achieving loose coupling and a balance between stability and flexibility – than as the general assurance of safety as a core value.

This is not to say that safety does not depend on people sharing a strong sense of the importance of safety. There is very convincing empirical evidence that safety climate, defined in terms of common perceptions of various elements of a company's safety-related activities, such as management commitment to safety, safety training, or safety as a promotion criterion (e.g. Flin *et al.* 2000; Zohar 1980), is related to safety outcomes like injuries or success of safety programmes (e.g., Hofmann and Stetzer 1996; Mearns *et al.* 2003; Neal *et al.* 2000; Zohar 1980, 2000). Safety climate is usually understood as one expression of safety culture, which, based on Schein's model of organizational culture, concerns deeply rooted assumptions about the world and what it takes to safely run an organization in this world (Grote and Künzler 2000; Guldenmund 2000; Pidgeon 1998). Assessing safety culture is far more complex than assessing safety climate, as these basic assumptions are often not conscious and cannot be reported on to the degree that espoused values and attitudes can be expressed, for instance, in survey instruments. Thus, safety climate assessments are often taken as proxies for evaluations of safety culture.

As was stated earlier, much of the research and practical work on safety culture and safety climate has had a fairly narrow focus on norms and values expressing the importance and quality of directly safety-related practices. The broader understanding of safety culture as a mechanism for coordinating and integrating organizational activities stressed by Weick (1987) and its consequences for assessing and improving safety culture has rarely been addressed (Pidgeon 1998; Grote 2007a). Given the particular relevance of this understanding for achieving a balance between stability and flexibility, it was taken as the starting point for developing a procedure concerned with assessing and designing safety management and safety culture from an uncertainty management perspective. Because the conceptualization of the relationship between autonomy and safety is crucial for moving from the predominant minimizing uncertainties approach in safety management towards a more balanced approach also incorporating coping with uncertainties, this relationship will be reviewed in considerable detail before the proposed assessment and design procedure is presented.

7.5 The Uncertain Relationship between Safety and Autonomy

At the beginnning of this chapter the relationship between safety and autonomy was discussed in terms of the debate between Perrow (1984) and the authors of the HRO studies (LaPorte and Consolini 1991; Roberts 1993). This debate is focused on whether organizations can achieve the balance between decentralized, autonomous action and centralized coordination required by the demands of both complex local interactions within and between subsystems of the organization and tight coupling between system components. All authors involved agree that local autonomy is necessary for the safe functioning of complex systems. However, Perrow argues that there is an incompatible demand for autonomy as well as for centralized coordination. The authors favouring the HRO-approach assume more optimistically that work systems can flexibly change between different modes of more or less (de-)centralized decision-making based on a strong culture, thereby resolving these seemingly incompatible demands.

The predicted beneficial effect of local autonomy as implied by the HRO approach has been one of the most contested in safety literature, especially regarding shop floor operations. In order to reassess the relevance of providing local autonomy in high-risk organizations, an overview of studies on the relationship between safety and autonomy is given below (see also Grote *et al.* 2003; Grote 2007a). As will be seen from this review, there is no unequivocal evidence for either a positive or a negative relationship, but there are mostly mixed results. Reasons for this will also be discussed, indicating mainly methodological issues that render a straightforward interpretation of existing evidence difficult.

There is substantial evidence supporting a positive relationship between autonomy and safety. In a recent meta-analysis of ten cross-sectional studies reviewing the relationship between different organizational and workplace factors and injury rates, autonomy – measured in terms of empowerment through, for example, encouragement of worker initiative instead of carrying out tasks based on management instructions – was among the factors frequently being linked to lower injury rates (Shannon *et al.* 1997). Also, in longitudinal studies comparing work accidents and unsafe acts before and after the introduction of self-managing teams, a positive relationship was usually found between autonomy and safety, though sometimes there was no relationship at all (for an overview Pasmore *et al.* 1982; examples of studies are Cohen and Ledford 1994; Trist *et al.* 1963, 1977). As another, more recent example, in a longitudinal study in two glass manufacturing plants where a safety campaign had been carried out, Parker and colleagues (2001) found a positive relationship between job autonomy and safe working, operationalized as self-reported rule compliance, which was mediated by organizational commitment.

There are a few older cross-sectional studies where a negative relationship between autonomy and safety was found by comparing levels of job autonomy in production companies with different work accident rates (*e.g.*, Hoyos *et al.* 1981; Saari and Lathela 1978).

Many studies have yielded inconclusive results, suggesting some positive and some negative relationships between different aspects of autonomy and safety. Using assessments of safety management quality by external auditors in six petrochemical plants as a measure of safety, Grote and Künzler (2000) found mixed evidence regarding different aspects of job autonomy measured through questionnaire data from operations and maintenance personnel. Higher perceived operational autonomy as well as higher order autonomy (*i.e.*, autonomy regarding the restriction of one's own operational autonomy, *e.g.*, through involvement in the definition of operating procedures and in organizational planning) were positively related to adequate handling of risks as judged by external auditors. However, autonomy related to the independent handling of disturbances – as one of the core requirements embedded in the coping with uncertainties approach to organization design – did not appear to be beneficial to system safety, as plants where this aspect was evaluated positively by plant employees were assessed particularly negatively regarding overall safety management by the external auditors.

Similarly mixed results were reported by Turner (2002). In a longitudinal study of the effects of introducing self-managing teams, he found a positive relationship between role enrichment and safety proactivity, but no relationship with safety compliance. This suggests that autonomy is only related to those safety-relevant behaviours that actually require some autonomy, whereas mere rule-following is not affected.

Another approach has been to study levels of job autonomy in relation to different elements of safety awareness. Research by Grote and Künzler (1996a, b) investigated a number of facets of the relationship between safety awareness and job autonomy. It was found, for instance, that higher perceived job autonomy was linked to a weaker emphasis on technology as a risk factor and to a stronger emphasis on the human as a risk factor. Similar results have been reported by Leplat (1987), indicating a link between autonomy and taking over (safety) responsibility.

Three meanings of safety as suggested by Kaufmann (1970) – personal shelteredness, technical system reliability, self-confidence – were used for the qualitative analysis of answers to open questions on the personal understanding of safety and safe behaviour in four chemical companies (Künzler 2001). Generally, safety and safe behaviour were predominantly linked to safety as personal shelteredness and technical system reliability. However, workers who perceived themselves as having higher levels of job autonomy were more likely to understand safety in terms of self-confidence also. When interpreting this finding, it has to be kept in mind that high self-confidence can have beneficial and detrimental effects on safety, as it may spur personal initiative and proactive behaviour, but also inadequate risk taking. This is supported by research showing a negative relationship between perceived control over a situation and risk perception (*e.g.*, Grote and Künzler 1996a; Jungermann *et al.* 1991).

7.5.1 Interpreting the Evidence

Several methodological problems with the current evidence have to be acknowledged. First, there are no commonly used operational definitions for either of the core variables, that is, autonomy and safety. Measures of autonomy have been, for instance, operational degrees of freedom, decision latitude regarding working conditions, or amount of task delegation, thereby sometimes stressing control, that is the influence on a given situation in order to reach goals, and sometimes autonomy, that is the self-determination regarding goals and the rules to be followed in achieving these goals (Grote 1997, 2004b). Safety, even when understood not only as occupational safety directly related to the health protection of employees, but more broadly as system or process safety, has often been measured by generic work accidents like bruises and cuts not necessarily related at all to process risks. Also, different types of safety-related behaviours have been looked at, *e.g.*, compliance-oriented versus proactive behaviour (Marchand *et al.* 1998). These diverse operationalizations could clearly impinge on the relationship between autonomy and safety. Moreover, autonomy might be linked more to process safety than to occupational safety, and might be detrimental regarding reactive indicators of safe performance like rule compliance, but positive regarding proactive indicators of safe performance like safety suggestions.

Secondly, there is a mix of longitudinal and cross-sectional studies. While the longitudinal studies on effects of increased autonomy show positive relationships between autonomy and safety more often and also allow for causal interpretations of these relationships, frequently they do not contain a proper control group and repeated measures designs checking for Hawthorne effects (Pasmore *et al.* 1982). The cross-sectional studies not only provide mixed evidence on the relationship between autonomy and safety, but also do not permit causal interpretations of beneficial or adverse effects of autonomy on safety. Summarizing therefore, the empirical evidence on the relationship between autonomy and safety is far from conclusive, leaving scientific and practical debate open to the influence of basic, often implicit assumptions about humans as risk and/or safety factors in complex systems.

In an attempt to advance the debate conceptually, it has been proposed recently that operational uncertainty might act as a moderator in the relationship between autonomy and safety, just as it does in the relationship between autonomy and performance (Grote *et al.* 2003; Wall *et al.* 2002). The more uncertainty there is, the more local autonomy and control is needed to cope with this uncertainty effectively and efficiently. This hypothesis follows the general contingency models for organizational design discussed in Chapter 3 and therefore may seem very plausible, but so far it has not been tested empirically. Some indirect support for the hypothesis has been provided by two studies in hospitals using a safety climate measure that included perceptions of the level of detail of procedures as a separate dimension (Katz-Navon *et al.* 2005; Naveh *et al.* 2005). The findings indicated that in this rather uncertain work setting, fewer treatment errors were related to a medium level of detail or more generally to the perception of a suitable level of detail. In the following final part of this chapter, a way forward is suggested which focuses more explicitly on the relationship between safety and uncertainty than has been the case in previous approaches to safety management.

7.6 An Uncertainty Management Approach to Safety Management and Safety Culture

As for organizational and system design in general, the basic contingency assumption is used as a starting point, that is high uncertainty requires coping, while little uncertainty can he handled via minimizing. However, from the preceding discussion two more specific assumptions regarding the assessment and design of safety management and safety culture can be derived:

1. The role of safety culture is different in the two approaches to uncertainty management. Safety culture is a crucial "soft" coordination mechanism within the competent coping approach to uncertainty management which helps to coordinate autonomous actors in achieving a superordinate goal. In the minimizing uncertainties approach, the role of safety culture is that of an additional coordination mechanism on top of the "hard" coordination through centralization and standardization.
2. Consequently, not only the appropriateness of the chosen approach to uncertainty management, but also the role of safety culture has to be reflected upon with respect to the contingencies an organization is faced with, above all the kinds and degrees of uncertainties to be handled.

Therefore, the assessment of safety culture should always be preceded by an analysis of the uncertainties an organization is confronted with and the appropriateness of the chosen ways for handling those uncertainties. Subsequently, a more focused assessment of safety management and safety culture should be undertaken to finally be able to draw conclusions on necessary measures regarding both the management of uncertainty and safety management. More concrete proposals for conducting these assessments are described in the following.

7.6.1 Analysis of Uncertainties and the Ways of Handling Them

In order to identify relevant uncertainties, the classification suggested in Chapter 2 can be useful, that is, to distinguish between contents, sources, causes, and potential consequences of uncertainties. Based on this, threats to system control by different actors and thereby to system safety, can be assessed. Subsequently, a rough overall assessment of the chosen type of uncertainty management should be undertaken, using the characteristics described in Figure 3.1 (see Chapter 3):

- How is planning done in the organization (prescriptive central planning versus planning as resource for situated local action)?
- How are planning and implementation linked (strict links via automation and standardization versus loose links through operative decision latitude and lateral coordination)?
- How are disturbances dealt with (more central planning and stricter links between planning and implementation versus local learning and adaptation)?

Moreover, indications of loose coupling should be investigated which would allow for a finer tuning of autonomy and central control depending on situational demands:

- Is the organization capable of switching between the two ways of handling uncertainties and how are the individual actors supported in making these switches?
- Is higher order autonomy used to provide actors with decision authority for setting standards for situations in which, on the operational level, adherence to fixed rules is essential?

Once the uncertainties as well as the current approach to uncertainty management have been identified, the fit between the two can be assessed, with the basic rule of thumb being that minimizing uncertainties will only work with low levels of uncertainties while coping with uncertainties is better for high levels of uncertainty (see Chapter 3).

As was stated before, this rule of thumb has only been tested for general performance so far, not specifically for safety performance. Most likely, some differentiation of this rule is needed. In order to support the assessment of the chosen ways of handling uncertainties, some more specific relationships between safety, uncertainty, and autonomy/control are suggested. For this purpose, the distinction between control, operative autonomy, and higher order autonomy (Grote 1997), between proactive and reactive safety behaviour (Marchand *et al.* 1998), and between process and occupational/personal safety (Grote 2007b; Hopkins 2009) are drawn on (see Table 7.1). The most important of these suggested relationships are:

- The contingencies are assumed to hold primarily for safety behaviour directly linked to process safety, while safety behaviours linked to occupational or personal safety, such as holding the handrail when walking on stairs, are probably not affected by the level of uncertainty (Hopkins 2009).

- Control as the means of directly influencing a situation in order to reach a goal is generally considered to have positive effects, but is particularly important with higher levels of uncertainty (Karasek 1979; Wall *et al.* 1996).
- Operative autonomy is particularly beneficial for proactive safety behaviour whereas more reactive safety compliance is not affected or might even be affected negatively, especially with low levels of uncertainty (Marchand *et al.* 1998; Turner 2002).
- Higher order autonomy through participation in rule making is considered always to be beneficial, independent of the level of uncertainty or the type of safety behaviour (Bourrier 1998).

Once an assessment of the uncertainties and their management has been completed, a more focused analysis of safety management and safety culture should follow, as outlined in the next section.

Table 7.1. Suggested relationships between autonomy, uncertainty and safety (adapted from Grote 2007a)

Types of autonomy		Uncertainty	
		Low	High
Local control (Influence on a situation in order to reach goals)	Low	Weak negative effect on all safety behaviour related to process or personal safety	Negative effect especially on all safety behaviour related to process safety
	High	Weak positive effect on all safety behaviour related to process or personal safety	Positive effect especially on safety behaviour related to process safety
Operative autonomy (Self-determination regarding goals and rules to be followed in achieving goals)	Low	Positive effect especially on reactive safety behaviour related to process safety	Negative effect especially on proactive safety behaviour related to process safety
	High	Negative effect especially on reactive safety behaviour related to process safety	Positive effect especially on proactive safety behaviour related to process safety
Higher order autonomy (Involvement in decisions restricting one's own operative autonomy)	Low	Negative effect on all safety behaviour related to process or personal safety	Negative effect on all safety behaviour related to process or personal safety
	High	Positive effect on all safety behaviour related to process or personal safety	Positive effect on all safety behaviour related to process or personal safety

7.6.2 Assessing Safety Management and Safety Culture

Plenty of instruments have ben suggested for the assessment of safety management and safety culture (Guldenmund 2007). The reason for suggesting yet another method goes back to the earlier discussion on safety culture, where it was argued that safety culture is often not understood in the context of culture as a mechanism for loose coupling. Because, from an uncertainty management stance, this is a particularly important perspective on safety culture, we developed a so-called socio-technical model of safety culture (Grote and Künzler 2000; see Figure 7.1).

We worked from a generic definition of safety culture as an organizational culture conducive to safety and stressed the importance of considering the make-up of the overall socio-technical system when evaluating safety culture. In addition to material characteristics of the safety organization and the immaterial norms and beliefs regarding safety, the model contains core elements of socio-technical system design and the underlying norms and values. Specifically, we argued for incorporating the capability for coping with uncertainties into the organization by furthering coordination and integration of self-regulated units, which control disturbances at their source autonomously. Relevant basic beliefs related to socio-technical system design are trust in human competence and in the reliability of humans as compared to technical systems.

The model claims that the following elements are crucial to safety in organizations:

- proactive integration of safety into organizational structures and processes (proactiveness);
- joint optimization of technology and work organization in order to enable coping with uncertainties (sociotechnical integration);
- consideration of both material and immaterial characteristics of an organization with respect to safety management and socio-technical design (value consciousness).

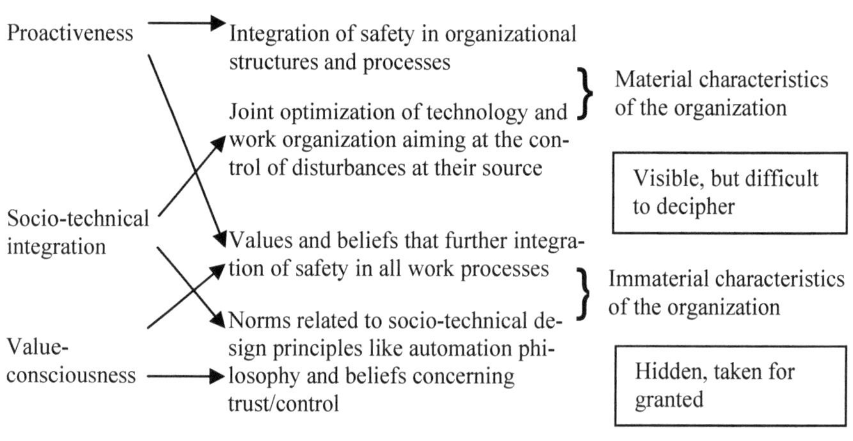

Figure 7.1. Socio-technical model of safety culture (from Grote and Künzler 2000)

The model aims to overcome shortcomings of previous models of safety culture by linking safety management, socio-technical systems design including the notion of controlling disturbances locally, and organizational culture. Safety culture is embedded in organizational culture, as not only directly safety-related norms and basic assumptions are considered, but also those related to system design. Furthermore, by addressing the control of disturbances, or uncertainties more generally speaking, the balance between minimizing and coping with uncertainties in the organization can be assessed. This is done regarding both the material characteristics of organizational design and the immaterial characteristics of organizational culture such as shared beliefs about human competence and the reliability of human and technical systems.

More concrete indicators for safety management and safety culture were derived from the model (see Table 7.2). From those, a questionnaire was developed to complement the information gained by interviews and plant visits conducted during safety management and safety culture audits, using the audit methods employed by Swiss Re – our partner in this endeavour – as a model (Müller *et al.* 1998). The questionnaire consists of three parts, (1) perception of safety management, (2) perception of management strategies and underlying values regarding both safety management and socio-technical system design, and (3) perception of the quality of job design.

Table 7.2. Indicators for assessing organizations based on the socio-technical model of safety culture (from Müller *et al.* 1998)

(I) Proactive promotion of safety
Strategic and operational safety goals for all company areas
Resource planning as commitment to the promotion of safety
Measurement of negative and positive safety indicators
Institutionalized process of continuous improvement
Adequate change management for evolutionary and radical change
(II) Socio-technically integrated system design
Task-adequate technical and organizational safety systems
Self-regulation in small control loops
Individual motivation through task orientation
Controllable technical systems
(III) Value-conscious actions
Individual and collective actions reflecting safety awareness
Reflection of the balance between centralized / decentralized autonomy and control
Participation of all concerned in decision-making and change processes
Questioning attitude regarding practices and guidelines

Table 7.3. Perceptions of safety and system design strategies (statements on the left reflect a minimizing uncertainties and on the right a coping with uncertainties approach)

Employees are bound to safety by strict control.	2.......1.......0.......1.......2	Employees are motivated for safety by information and interesting tasks.
Employees do not have to assume safety-critical operations.	2.......1.......0.......1.......2	Employees are qualified to actively enhance operational safety.
Independent working decisions by plant personnel are avoided because they are error-prone.	2.......1.......0.......1.......2	Plant personnel are encouraged to independently carry out their work responsibilities.
During process upsets plant personnel need to follow procedures and specialists' instructions.	2.......1.......0.......1.......2	During process upsets it is important that plant personnel are able to make their own decisions.
Instructions are not meant to be questioned.	2.......1.......0.......1.......2	A questioning attitude towards instructions is promoted.
Procedures are developed by specialists.	2.......1.......0.......1.......2	Procedures are developed in close cooperation with plant personnel.
Technology is used to replace people.	2.......1.......0.......1.......2	Technology is used to support people in fulfilling their tasks.
Plant personnel may not intervene in automated processes in order not to jeopardize safety.	2.......1.......0.......1.......2	Plant personnel can intervene in automated processes to ensure quality and safety of production.
Safety issues have to be dealt with by following the chain of command.	2.......1.......0.......1.......2	Safety issues can always be discussed with superiors.
Safety is ensured by specially trained and assigned safety officers.	2.......1.......0.......1.......2	Safety is everyone's responsibility, and safety officers provide support.
Near misses are not discussed.	2.......1.......0.......1.......2	A lot is learned from near misses.
Safety goals are continuously adapted to comply with the newest regulations.	2.......1.......0.......1.......2	Safety goals are pursued proactively and on the company's initiative.

While the first and third part are very similar to other survey instruments for the assessment of safety management and safety climate or culture (Flin *et al.* 2000), the second part is rather unique. Statements are contrasted that reflect the minimizing versus coping with uncertainties approach for both safety management and socio-technical system design more generally (see Table 7.3).

The questionnaire has been applied extensively in insurance audits in the petrochemical industries (Grote and Künzler 2000; Grote 2008a). One case was particularly interesting, because we had the opportunity to conduct an audit twice in the same company in order to evaluate changes in safety-related perceptions in the course of a reorganization programme. In the first survey, the very positive results could be linked to a strong and shared effort to improve safety management. Views two years later seemed more sober as the plant had just overcome some conflicts between management and unions over the reduction of central safety staff and delegation of safety responsibility to line management. Moreover, employees were less sure of the general direction safety efforts would take under a recently appointed new plant management. The tension between management and the shop-floor workers was especially evident in the far more critical evaluations by workers compared with supervisory personnel. Interestingly, these differences existed mainly in the assessment of material indicators of safety management, and not so much in the assessment of safety and design strategies. For those, responses were fairly similar for both groups as well as between the two measuring times.

These findings indicate that the second part of the questionnaire referring to safety and design strategies is able to capture more general and enduring aspects of safety management, which might be considered expressions of culture while the first part of the questionnaire referring to safety management is related more to the immediate situation in a company. However, the questionnaire is not as such intended to measure culture, as no questionnaire could ever capture the complexities of the unconscious beliefs and assumptions which make up an organization's culture, but it is considered as a basis for discussing organizational design issues and the more or less shared values behind them. Therefore, as part of the audit instrument for which the questionnaire was developed, there is always a feedback meeting with members of the audited organization during which the audit results are discussed. The discussion of responses to the questions on safety and design strategies in particular have been found to support an increased consciousness of existing norms and values and their interaction with the organization's safety efforts. Overall, the questionnaire helps in broadening and detailing the information gained in safety management audits, allowing auditors to make explicit and to test diagnoses of safety culture that before were predominantly "gut feelings". In addition, the companies audited usually reacted very positively as they obtained a more informed outside view of their own situation which could serve as a basis for more constructive criticisms and suggestions for improvement.

More recently, the questionnaire has been extended to include the explicit assessment of change management in relation to an organization's safety standards (Grote 2008a). Four factors were found to be important for good change management: respect for employees, procedures for organizational change, transparence of the change process, and a vision as guiding principle. This new instrument allows to evaluate changes such as the ones described in the case example above, more systematically.

7.6.3 Adjusting Safety Management and Safety Culture in Light of Adequately Managing Uncertainty

The overriding assumption is that a strong, shared culture is more important in organizations operating under the coping with uncertainties approach than under the minimizing uncertainties approach, given that the chosen approach fits the coordination requirements resulting from external and internal uncertainties. In the latter approach, organizational and technical standardization act as depersonalized "hard" coordination mechanisms, supplemented by a command and control type leadership as a personalized, but still rather "hard" form of coordination. In the case of routine work, such coordination will be adequate, with "softer" coordination through participative leadership, reciprocal team interaction and cultural norms and values mostly needed for keeping people involved and motivated, but not for coordinating their activities as such. This is not to say that shared beliefs about the importance of safety and a strong commitment to policies, standards, and behaviours ensuring good safety performance are not beneficial under those circumstances. Of course they are, and, on the other hand, beliefs that further complacency or violation of safety standards are clearly detrimental. But the point is that these cultural norms act as additional and not as the main force for coordinating work activitites.

When in an organization, at least some of the time, coping with uncertainties is required due to high levels of uncertainty, "hard" coordination is insufficient or even detrimental because feedforward control through standards, technical algorithms, and directive leadership is not possible. Instead, feedback control is needed, which can be achieved through "soft" forms of coordination, that is, reciprocal team interaction and participative and empowering leadership based on a shared understanding of task requirements and interdependencies. General standards may help to orient behaviour in this case, but cultural norms and values will be needed to interpret and adapt standards according to situational demands. The intricate adaptations happening in incident command systems described earlier (Bigley and Roberts 2001) are an excellent illustration of this necessity.

In both cases results from the assessments of safety management and safety culture should, of course, be taken very seriously and needs for improvements addressed, but it is argued here that the urgency of improving culture is much higher in the second case. When in a decentralized organization, which operates at least some of the time in a coping with uncertainties mode, shared beliefs and norms as guides for coordinated work activities are missing, it could be devastating.

If, however, the assessment shows the fit between uncertainties and uncertainty management to be inadequate, the material workings of the organization need to be changed in order to establish a more appropriate form of uncertainty management, before safety management and safety culture are addressed in more detail. In order to decide on necessary changes, the design process suggested in Chapter 3 can provide some guidance. Costs and benefits of reducing, maintaining or increasing uncertainties should be discussed and decisions made accordingly along with designing specific measures for managing the resulting uncertainties. The railway case presented in Chapter 3 is a good illustration because safety is very directly affected by the different ways of handling uncertainties discussed there. Another example stems from analyses done at Swiss Re, which show that US petrochemical plants have proportionally more accidents than, for instance, Russian plants. One explanation offered for this rather unexpected result is that in Russian plants there is a strong awareness of uncertainties related to the age and overall poor maintenance of plants, which are counteracted by reducing operational uncertainties as plants are never run at or even near their full capacity. In the US, on the other hand, uncertainties are deliberately increased beyond good measure by trying to run the plants as close to full capacity for as long as possible.

Change efforts should tackle the manifest problems in organizational design first, before worrying about safety culture. However, current culture may get in the way of establishing better organizational practices (see Example 7.1 for an example). Commonly, change meets resistance that may partially be founded on well-argued objections, but is always linked to cultural norms and values as well which support the current state of affairs. These norms and values will then also have to be reflected upon and gradually modified along with the changes in organizational design. If culture is to serve as a "soft" form of centralization in an otherwise decentralized organization, particular attention needs to be devoted to building and maintaining a strong culture.

Example 7.1. Uncertainty management in a nuclear power plant

In a safety management analysis in a nuclear power plant perceptions of safety and system design strategies were studied with the particular focus of furthering coping with uncertainties. The generally expressed belief in this plant was that much decision latitude is provided for operative personnel because the need for flexibility is explicitly acknowledged. This opinion was also voiced in a number of interviews carried out with members of line management. In a survey on safety and system design strategies, it became obvious that employees in operations did not perceive strong support for the coping with uncertainties strategy, while management and employees in the mechanical and electrical departments did perceive this strategy to be the dominant one. Also, analyzing the responses for each item separately, it could be seen that the lowest general support was expressed for items very directly linked to operative degrees of freedom (plant personnel can intervene in automated processes to ensure quality and safety of production; during process upsets it is important that plant personnel are able to make their own decisions) and for higher order autonomy through participation in rule making. These results were discussed in terms of that plant's relationship to the regulator. It was found that the emphasis on flexibility and degrees of freedom was geared more towards protecting the organization's independence from increasing regulatory pressures than towards furthering internal coping with uncertainties. The strong opposition to the regulator even hindered a systematic internal reflection on approaches to handling uncertainties. In this particular case, openness within the management team for systematically assessing the fit between uncertainties and the chosen ways of managing them will have to be created first, before any decisions can be taken regarding possible changes in organizational practices.

However, "safety culture engineering" as suggested by Reason (1997) will rarely be successful. Instead culture change is a process that can only partially be directed, but rather is influenced by introducing manifest changes in organizational structures and processes. Additionally, continuous reflection upon practices and underlying basic assumptions and their appropriateness for providing orientation to autonomous actors in highly demanding environments is needed. Safety culture is mostly the consequence of a well-managed change process focusing on improving the management of uncertainties, and to a much lesser degree, affected by deliberate attempts to create a new set of norms and values.

7.7 Chapter Summary

In this chapter, the management of safety was discussed from the perspective of managing uncertainties. Safety has long been managed mainly by attempting to minimize uncertainties following the basic understanding of safety as minimal acceptable risk. However, minimizing uncertainties implies losing the organizational flexibility needed for resilient responses to disturbances. Therefore, newer approaches to safety management aim to combine different ways of handling uncertainties into strategies that improve the balance between stability and flexibility in line with the notion of loose coupling. Safety culture is seen as important for achieving loose coupling because culture can serve as a "soft" mechanism for coordination and integration in a formally decentralized organization. As there is much reluctance to the coping with uncertainties approach in high-risk organizations, in particular due to the basic assumption that individual autonomy furthers errors, the empirical evidence on the relationship between safety and autonomy was discussed in more detail. It was shown that, to date, there is no complete understanding of this relationship, but that there is good reason to assume that uncertainty is an important moderator between autonomy and safety, just as it is between autonomy and performance more generally. Consequently, safety management and related design questions regarding the distribution of autonomy and control in an organization should be approached from the perspective of managing uncertainties. A procedure for evaluating and designing safety management and safety culture was described. This procedure emphasizes the importance of assessing the fit between uncertainties and their management and of changing organization design to improve the fit if needed. Change efforts should mostly worry about culture in terms of current belief systems getting in the way of establishing better organizational practices. However, in order to promote loose coupling, attention also needs to be paid to advancing culture as a mechanism for integration and coordination.

8

Managing Uncertainties in Working Life

So far, a predominantly organizational perspective on the management of uncertainties has been taken, focusing on uncertainties that are directly related to organizational and work processes. In this chapter we will switch perspectives and take on the view of the individual actors in organizations as they try to cope with uncertainties in their everyday work, as well as in their working lives more generally. As it is people that accomplish an organization's goals, this change in perspectives is very instructive. The individual-focused perspective will be translated back into an organizational view at the end of the chapter discussing consequences for human resource management.

Much of the management literature aims to support strategic decision-makers in handling uncertainties, including – at least indirectly – their personal uncertainties related to performing well and advancing their careers. One may even argue that many strategic decisions in organizations are motivated by personal attitudes towards uncertainty, as was discussed in Chapter 3 in terms of the influence of belief systems on managing uncertainties. From that perspective, this whole book is about those individuals in organizations who are in charge of strategic decision-making and how they handle uncertainties. In the following, the focus will be shifted to the "ordinary" employee, whose job it is to deal with operational uncertainties in everyday work processes, and who also faces uncertainties more generally in his or her employment relationship.

Firstly, job design will be discussed because it strongly influences both the day-to-day uncertainties employees have to deal with in their work and the resources they have available to manage these uncertainties. Subsequently, uncertainties that affect the employment relationship more broadly will be focused upon. Specifically, increasing work flexibility and its effects on psychological contracts, new forms of career and employability requirements will be laid out. Finally, consequences for human resource management will be described and some specific measures for improving the management of uncertainty in employment relationships proposed.

8.1 Uncertainties in Everyday Work

Organizational design always affects job design, which in turn, determines the uncertainties each indiviual worker has to face as well as the organizational resources he or she has available to handle these uncertainties. From the outset of this book, this individual perspective has been implicit though the control framework chosen to define uncertainties. The general discussion of system design as well as the topic-specific considerations in the subsequent chapters all aimed at designing conditions for optimal control. The actors who will be subjected to these conditions and presumably should be able to actually control work processes have only been discussed in passing, though. From a systems as well as an actor perspective, crucial prerequisites for control are system transparency and predictability of system behaviour based on knowledge of cause-effect relationships, and possibilities for action (Brehmer 1992; Sutton and Kahn 1987). The previous chapters described how these prerequisties can be established in different work settings. What is still missing, is a discussion of how actors themselves perceive and act upon these conditions. This has been dealt with extensively in the job design literature, so just a summary of that literature is provided here.

The core tenet of job design is the assumption that people can be and even should be intrinsically motivated to do their work. Intrinsic motivation implies motivation stemming from the activity performed itself and whatever benefits result for the individual in the course of performing the activity, such as developing and demonstrating certain knowledge and skills, experiencing socially rewarding exchange with other workers, or nurturing intellectual curiosity. Several theories have been developed to explain under which conditions intrinsic motivation evolves and what the underlying psychological processes are (*e.g.*, Deci and Ryan 1985; Heyman and Dweck 1992). From these theories, job design has taken the pragmatic notion that complete, varied and meaningful tasks in combination with performance feedback and autonomy – understood as freedom in scheduling work and determining how to carry it out, which represents a mix of autonomy and control as defined in Chapter 3 – are crucial ingredients of jobs that further intrinsic motivation (*e.g.* Hackman and Oldham 1976).

Another important stream of research has been concerned with defining conditions for well-being at work, focusing in particular on stress at work and how it can be avoided. Stress has been defined as the subjective assessment that demands posed in a situation exceed the individual's resources for coping with them (*e.g.*, Lazarus 1999). Probably the most influential stress model in job design has been Karasek's job demands-job control model (Karasek 1979; Karasek and Theorell 1990), which states that the opportunities for control at work must match the temporal and intellectual demands made on the worker in order to keep stress at an acceptable level. Control in this model has been defined similarly to the definition used in this book as a means of influence in combination with transparency and predictability of the situation. Interestingly, support for the theory was found especially when job control was measured in those terms instead of using more generic measures of decision latitude (Wall *et al.* 1996).

Taking the existing evidence on the importance of job autonomy and control – often using these two terms interchangeably despite their different meanings (see Chapter 3) – much of the research in job design has aimed at corroborating assumptions about relationships between job characteristics and individual performance and well-being (Parker and Wall 1998). Though there is strong support for control (as a direct means of influencing work processes) and empowerment (as delegated autonomy on decisions regarding goals and ways to reach these goals) being crucial for performance, satisfaction and health, the evidence is not unequivocal. Recently, Wall and colleagues (2002) have presented an extensive review of the literature across different research streams such as human resource management, quality management and human factors. They argued that uncertainty needs to be included as a moderator in the studied relationships, that is, job autonomy and control lead to better performance particularly under conditions of high operational uncertainty because coping with uncertainties is supported. This leads us back to the main thrust of this book.

8.2 Uncertainties in Working Lives

Beyond the uncertainties experienced in trying to control different kinds of work processes, individuals are also confronted with uncertainties associated with their broader employment relationship. The framework used for defining uncertainties in this book also fits these uncertainties if one shifts the attention away from decisions and actions directly related to particular work processes towards more general work-related decisions and actions concerning organizational commitment, intention to leave the company, or involvement in training and further education. The overall objective also shifts then from maintaining efficient and effective work processes to engaging in a successful career. Particular uncertainties concern, for instance, the degree of job (in)security, the availability of training opportunities or the chances for employment on the internal or external labour market. Causes of uncertainty can be internal, such as major organizational restructuring, or external, such as changing demand for certain occupations.

In recent years, much has been written about increasing uncertainty in employment relationships, especially with respect to changes in the psychological contract – defined as implicit assumptions about employer inducements and employee contributions in the employment relationship – and in career patterns (*e.g.*, Arthur and Rousseau 1996; Rousseau 1995). Increasing economic pressures have led many companies to undertake major change programmes to further flexibility in their internal operations. For more and more jobs, there is the option of as well as the requirement to work anytime and anyplace. Most compensation systems include flexible pay-for-performance schemes. Job assignments and job descriptions become broader and change more often and increasing numbers of employees work with fixed-term contracts and less protection against job loss. With respect to human resource management practices, Reilly (1998) has distinguished between five types of work flexibility:

- locational (work being carried out at different locations, *e.g.*, working from home),
- temporal (open time arrangements, *e.g.*, flexitime),
- functional (changing/expanding job descriptions),
- financial (variable pay, *e.g.*, pay for performance), and
- numerical (open forms of employment, *e.g.*, contract work).

Through these different types of flexibility, the employment relationship becomes less legally prescribed, leaving more room for actively designing the relationship for both employer and employee. It has been argued that the economic pressures underlying the trend towards more flexibility also lead to more economically defined relationships, for which however, there is only mixed empirical support as will be discussed later. Undoubtedly, flexibility increases the uncertainty in the employment relationship for both parties. Employees are confronted with vague requirements for their constant availability and continuous learning. Employers may find that employees are less willing to invest in the relationship in terms of extra-role activities and being good "organizational citizens".

These flexible work arrangements have affected the employment relationship in a number of ways, for example, by increasing the use of contingent labour (*e.g.*, Barley and Kunda 2004) and causing career paths to cross organizational borders more frequently (*e.g.*, Arthur and Rousseau 1996). It is assumed that careers characterized by major occupational reorientations, including changes of employer and occupation and periods of unemployment, will become prevalent (*e.g.*, Arnold 1997; Arthur *et al.* 1999). Demands for handling uncertainties in the employment relationship will increase, both for the individual and the organization. These demands will be discussed under three headings: changing psychological contracts, new career forms, and employability.

8.2.1 Changing Psychological Contracts

Before changes in psychological contracts in today's organizations are described, a theoretical framework for understanding psychological contracts is presented, based on a social exchange perspective.

Social exchange as basis for the employment relationship. The relationship between employee and employer can be understood purely economically as an exchange of work for money. This economic exchange relationship is also the core of legal work contracts, which usually specify additional characteristics of the employment relationship such as working hours, handling of sick-leave, and conditions for terminating the relationship. Beyond the formally determined exchange there are always elements of social exchange in employment relationships regarding company inducements and employee contributions that are not written down and potentially not even consciously influenced by the exchange partners (Coyle-Shapiro and Conway 2004). Frequently, social exchange also concerns more intangible benefits compared to a purely economic exchange, such as loyalty, commitment or social support. While reciprocal obligations and means for securing compliance are explicit in legal contracts, social exchange is more vague and trust in partners' compliance has to be built slowly.

An important condition for social exchange is reciprocity (Gouldner 1960; Coyle-Shapiro and Conway 2004). Reciprocity can be achieved through heteromorphic and homeomorphic exchanges. The latter involves exchanging the identical content under identical circumstances, for instance, being helped by a colleague at work to meet a deadline and later helping that colleague in the same way when he or she is under time pressure. The former is an exchange of equally valued, but different things, taking into account the different needs of the exchange partners. I might be helped by a colleague who rarely needs help herself, but appreciates being publicly praised for her competence in return. The value of a benefit received also depends on how much in need the receiving party is and how costly, voluntary, and altruistic providing the benefit is on the side of the donor. Not reciprocating immediately also furthers the relationship because it is then seen less as purely economic.

One very basic characteristic of the employment relationship in comparison with other kinds of social exchange is the power differential in the relationship. By signing a work contract, employees agree to provide the employer with more authority over the relationship than they themselves obtain. For the quality of the ensuing relationship, the organization's ways of using or abusing this power differential over the course of a potentially long relationship with many changes in external and internal conditions is crucial. In the context of increasing uncertainties due to increasing flexibility in the employment relationship, this power differential becomes particularly critical. Overall, there is a tendency to shift uncertainties to the employees, who have to take on responsibility for career management, skill development, and employability beyond their directly job-related responsibilities, as organizations abandon job security and long-term loyalty in favour of short-term adaptations in the number and kind of employees needed (*e.g.,* Arnold 1997; Greenhouse 2008; Raeder and Grote 2001).

Because employment relationships concern parties of unequal power the perceived justice of the exchange is crucial for relationship quality. Moreover, it has been found that perceived justice is particularly important under conditions of high uncertainty where trust in other actors needs to be built (Lind and Van Den Bos 2002). Three kinds of justice can be distinguished (Folger and Cropanzano 1998):

- *distributive justice*, which concerns the distribution of material and immaterial goods between the partners in the exchange relationship, but also the comparison with individuals and groups external to the relationship;
- *procedural justice*, which concerns the procedure chosen to determine the distribution of goods, for instance, the criteria used to decide on the (non-)equity of the distribution;
- *interactional justice*, which concerns communication about the procedure and results of the distribution decisions, for instance, the transparency of information provided.

When studying or designing social exchange relationships at the workplace, a basic issue is the question of who the partners in the exchange actually are. On the employee side this is easy to answer; on the employer side it is much more difficult. Is it the formal organization, is it top management, is it the Human Resources department, is the direct supervisor, or is it a conglomeration of all people in the organization? It seems that, depending on which element of the exchange relationship one focuses on, the answer can be any of the above.

Cropanzano and colleagues (2002) found, for instance, that for the perception of procedural justice, upper management and organizational policies are most relevant, while for the perception of interactional justice, the relationship with the direct supervisor is crucial.

Growing importance and vulnerability of psychological contracts. With increasingly open and adaptable legal work contracts, the importance of the social exchange relationship increases as a potential support in handling uncertainties. In employment relationships, the basis for social exchange can be understood as an unwritten contract which entails reciprocal obligations and expectations beyond those specified in the legal work contract. These unwritten contracts have been studied mostly under the term "psychological contract" (*e.g.,* Kotter 1973; Rousseau 1989). It is assumed that psychological contracts can help reduce uncertainties in the employment relationship; at the same time, given drastic and fast changes in organizations, these unwritten contracts have become more vulnerable (Rousseau 1995).

In line with this general understanding of its growing importance and vulnerability, researchers studying the psychological contract and its antecedents and consequences have focused in particular on contract breaches as perceived by employees. A "perceived breach" (Morrison and Robinson 1997) happens when employees perceive a discrepancy between their expectation for how well their employing organization should perform certain obligations and how well it actually fulfils them. Contract breach has mostly been studied without distinguishing between the different contents of expectations, but rather by looking at the overall extent of unmet expectations and their effects. For example, negative relationships between breach and job satisfaction and organizational commitment and positive relationships between breach and intention to leave the organization have been reported (*e.g.,* Cassar 2001; Conway and Briner 2002; Coyle-Shapiro and Kessler 2000; Robinson and Rousseau 1994).

When different contents of expectations are distinguished, there is often some reference to characteristics of a traditional contract such as loyalty, upward mobility, and job security, and to characteristics assumed to be part of a new contract such as performance orientation, lateral mobility, and competence development (*e.g.,* Hiltrop 1995; Raeder and Grote 2001). Another distinction found frequently in the literature is between relational and transactional contracts, where transactional contracts concern specific exchanges of directly economically relevant goods such as amount of pay for a specified level of performance and relational contracts are broader and more subjective involving exchanges of socio-emotional, value-based resources (Conway and Briner 2005; Rousseau 1990).

A number of studies have been carried out in order to test the assumption that psychological contracts have changed towards more transactional contracts comprising "new" contents such as performance orientation and training opportunities. Frequently, it is found that employees still value and expect contents of the "old" relational contract which focus on loyalty and security and that also, many employers still consider these to be important, while at the same time advocating more employee responsibility for employment and development (*e.g.*, Raeder and Grote 2004; Guest and Conway 2004). There seems to be some understanding amongst both employees and employers that the employment relationship and, more specifically, the psychological contract needs to be redefined in order to take into consideration new uncertainties, but only few examples exist where a new psychological contract has been established successfully. Mostly, these examples concern New Economy companies with highly skilled, younger employees who are willing to take on the new uncertainties in exchange for more freedom and challenging opportunities for career and personal development (*e.g.*, Raeder and Grote 2004).

It has even been argued that job security should remain a crucial element in successful employment relationships given the fact that it is still highly valued by particularly successful companies (Pfeffer 1998; Pfeffer and Veiga 1999). These successful companies have very selective hiring practices, but offer those employees they eventually do hire an "old" contract based on security and loyalty, while at the same time stressing performance orientation and skill development, which are considered part of the "new" contract. Google is still too young and too successful to have had the need to prove they can provide inducements of security and loyalty to their employees. Otherwise, they very much follow the pattern identified by Pfeffer, that is, selective hiring, empowerment, extensive training opportunities, performance-oriented pay, high availability of business information, and low status differences.

Individuals need to balance their need for security, development, and autonomy in order to shape uncertainties in the employment relationship according to their coping capabilities. Organizations, on the other hand, also have to evaluate the costs of employees leaving due to insufficient offers of security and internal development in relation to the benefits created by fast adaptation to changing patterns of required skills by hiring new people and laying off the current employees. A simple means of evaluating the fit between expectations, employee contributions and employer inducements is to include a measure of the psychological contract in employee surveys or in annual performance reviews. Table 8.1 describes such a measure successfully used in research as well as in practical interventions (Wittekind *et al.* 2006). In this measure elements of company inducements and employee contributions were chosen that relate to the distinction of the traditional versus the new psychological contract as discussed earlier. However the particular elements included could be adapted to whichever specific issues are of importance in a given company.

As has been pointed out by Kotter (1973) and more recently by de Vos and colleagues (2003), hiring and initiation into the company are crucial for establishing the psychological contract. It is therefore very important to communicate expectations from the start openly, both on the side of the employee and the employer. An interesting example is Novartis, where company inducements have been defined very explicitly in terms of an employer value proposition and are used consistently during recruiting, hiring, and initiation in order to support the development of realistic employee expectations. This example will be described in more detail at the end of the chapter when measures for managing the psychological contract are discussed (see Example 8.1).

Table 8.1. A measure for the psychological contract (from Wittekind *et al.* 2006)

I expect from my employer…						My employer offers…				
	Disagree	Rather disagree	Partly agree/ disagree	Rather agree	Agree	Disagree	Rather disagree	Partly agree/ disagree	Rather agree	Agree
Interesting work	❑	❑	❑	❑	❑	❑	❑	❑	❑	❑
Opportunities for taking on responsibility	❑	❑	❑	❑	❑	❑	❑	❑	❑	❑
Job security	❑	❑	❑	❑	❑	❑	❑	❑	❑	❑
Opportunities for using existing skills	❑	❑	❑	❑	❑	❑	❑	❑	❑	❑
Opportunity for personal development	❑	❑	❑	❑	❑	❑	❑	❑	❑	❑
Loyality	❑	❑	❑	❑	❑	❑	❑	❑	❑	❑
I offer my employer…						My employer expects of me…				
	Disagree	Rather disagree	Partly agree/ disagree	Rather agree	Agree	Disagree	Rather disagree	Partly agree/ disagree	Rather agree	Agree
Taking responsibility	❑	❑	❑	❑	❑	❑	❑	❑	❑	❑
Promoting my company's image	❑	❑	❑	❑	❑	❑	❑	❑	❑	❑
Identification with my company	❑	❑	❑	❑	❑	❑	❑	❑	❑	❑
Willingness to adapt to changing demands	❑	❑	❑	❑	❑	❑	❑	❑	❑	❑
Goal and performance orientation	❑	❑	❑	❑	❑	❑	❑	❑	❑	❑
Developing knowledge and experience on one's own initiative	❑	❑	❑	❑	❑	❑	❑	❑	❑	❑

8.2.2 New Career Forms

Along with the demands and opportunities created by more flexible employment relationships, there are changes in individual career paths and in individual and organizational career management. Just as there is talk of the new psychological contract, there is also the assumption of the "new career" (*e.g.,* Arthur *et al.* 1999). In career management literature these changes have given rise to concepts such as the boundaryless career (Arthur and Rousseau 1996) and the protean career (Hall 2002). Common to these concepts is the assumption that careers involve more self-directed job mobility and a shift in the employment relationship from long-term loyalty and employer-managed careers to a focus on competence development, employability and self-managed careers, with corresponding changes in the psychological contract (*e.g.,* Eby *et al.* 2003).

In recent years, many studies have addressed these new forms of career, often however, without providing empirical evidence on the prevalence of these career forms and their antecedents and consequences. Most accounts have been qualitative in nature, based on in-depth case studies of individuals with discontinuous careers (Arthur *et al.* 1999; Ibarra 2003; Barley and Kunda 2004). The general gist of these analyses is that career patterns have become more diverse, creating both opportunities and threats to people's lives.

The few studies that have used a broader database have usually focused on career orientations as general expressions of desired forms of career. In these studies it was found that new forms of careers are not (yet) widespread and often, coincide with negative outcomes such as low job satisfaction, high intention to quit the organization, and low organizational commitment. Distinguishing between independent, traditional, and disengaged career orientation, Guest and Conway (2004) found that for a representative sample of British employees, 25% of the respondents reported an independent career orientation and at the same time, were the least satisfied with their job situation. On the other hand, 36% followed a traditional career orientation and overall, expressed the highest level of satisfaction. Similar results were obtained in two consecutive studies with representative samples of employees in the German-speaking part of Switzerland (Wittekind *et al.* 2006, Gerber *et al.* 2007). Close to 20% of the respondents expressed an independent career orientation, which was also the best predictor of the intention to quit the job.

As new forms of careers are less bound to one organization and to success and upward mobility in that organization, new ways of conceptualizing career success have been discussed. Several authors have argued that subjective indicators of success, such as personal career satisfaction and general self-esteem built on psychological success, become more important compared to objective indicators, such as number of promotions and pay (*e.g.,* Arthur *et al.* 2005; Hall 2002; Mirvis and Hall 1994). From this perspective, the results may have to be reinterpreted regarding low satisfaction and commitment and high intention to quit for employees with an independent career orientation. Career success for these individuals may be expressed less in a "successful" employment relationship, but more in terms of fulfilling personal needs for change and development across sequences of employment relationships.

It might be argued that people who value flexibility and independence in their career, might be better able to handle the uncertainties inherent in more flexible employment relationships. Only partial support for this assumption was provided by Grote and Raeder (2009) by means of a complex combination of qualitative and quantitative data collection and analysis of two samples of individuals with either continuous or discontinuous careers working in organizations with different kinds and degrees of flexible work arrangements. They were able to distinguish two types of less traditional career identity, which they termed self-determined and critical-flexible. The self-determined type, which was the least frequent type and most commonly found in the more flexible organizations, was characterized by high subjective career success in terms of general self-esteem and satisfaction with the job situation and low objective career success in terms of hierarchical level in the organization. The critical-flexible type, which contained the majority of individuals with discontinuous careers, was characterized by both low subjective and objective career success. Interestingly, employees with career identities resembling traditional forms of career were characterized by more objective and subjective career success than employees with the two newer forms of career identity.

Analyzing the perceived psychological contract for the different career identity types, an astonishing finding was that the individuals in the self-determined type evaluated the offer of job security in the psychological contract more positively than both the critical-flexible type and the job-centred type, which was one of the two more traditional types identified. Only the continuous type, as the type with the most traditional career patterns, evaluated job security even more positively. Furthermore, the critical-flexible type displayed the largest negative discrepancy in the psychological contract between personal flexibility offered and the perceived flexibility demands by the company. In order to interpret this rather intriguing result further, more knowledge would be needed on the particular kinds of flexibility demanded and offered in line with recent research on "idiosyncratic deals" (Hornung *et al.* 2008). But even as it stands, the result points to more complex relationships between career orientations and the handling of uncertainties in the employment relationship than previously assumed.

In a study by Gerber and Grote (2007), the moderating role of career orientations on the relationship between psychological contract breach and work outcomes was examined. It was found that different content areas of unmet expectations had a specific impact on employees' work attitudes depending on their career orientation. The relationship between the perceived breach regarding elements of security and retention and the intention to quit was particularly strong for traditional-loyalty oriented employees. A traditional-promotion oriented career orientation moderated the relationship between unmet expectations referring to participation and commitment. A perceived breach regarding support for career and skill management led to particularly high intention to quit for employees pursuing an independent career orientation. The study thus confirmed that reactions to a breach of the psychological contract depend on employees' career orientation. Employers should therefore carefully take into account employees' career orientation when managing employment relationships by providing human resource and career management practices.

Drawing on Gouldner's (1957) distinction between local and cosmopolitan role orientations, Grote and colleagues (2007) again analyzed reactions to mismatches for different elements in the psychological contract. While locals are interested in following a career within one organization, showing loyalty to that organization, having an internal reference group orientation, and little commitment to their profession, cosmopolitans are strongly committed to their profession, have an external reference group orientation and show little loyalty to their organization. Consequently, it was assumed that locals would react more strongly to mismatches concerning "traditional" elements of the psychological contract such as loyalty and security; and cosmopolitans would react more strongly to mismatches in "new" elements such as competence development and performance orientation.

The results obtained from a sample of roughly 400 Swiss employees confirmed the findings of much of the previous research, namely that perceived breach of the psychological contract is related to a number of negative outcomes. From the findings, it appeared more specifically, that for all respondents job satisfaction was negatively influenced by any kind of breach. Perceived breaches regarding cosmopolitan elements predominantly affected the intention to quit and commitment for individuals with a cosmopolitan orientation. Individuals with a local orientation did not seem to be bothered much by mismatches concerning this "new" form of psychological contract. Perceived breaches regarding local elements were generally related to negative outcomes, indicating that mismatches in the more traditional local contract have negative effects on everything for everybody. These results may be specific to Switzerland where permanent employment contracts and long tenures are still the norm.

Overall, the empirical evidence suggests that careers are changing slowly and that newer forms of career are not necessarily promoted much in organizations. Many employees still look for security in their employment relationship, be it in the more traditional employment setting in Switzerland (Wittekind *et al.* 2006) or in the more "New Economy" oriented Great Britain (Guest and Conway 2004). A comparative study in Switzerland and the United Kingdom has again confirmed the finding that traditional career orientations are still the most widespread amongst the workforce and, moreover, also the existence of mixed types of career orientations (Gerber *et al.* 2008). It was shown that in three samples taken from culturally different regions, different types of career orientations can be identified. The finding that country- and even region-specific career orientations exist, and the understanding that culture affects career orientation, suggests a differential approach to the management of uncertainties regarding job security and career opportunities. Unfortunately, comparable data are not available for other countries and economic systems. If people are to be prepared for handling more uncertainty in their employment relationships, much still needs to be done.

8.2.3 Employability

One of the key factors discussed as a support for individuals in dealing with uncertainty in their employment situation, is the concept of employability. According to the most straightforward definition, employability concerns "an individual's chance of a job on the internal and/or external labour market" (Forrier and Sels 2003, p. 106).

Several models have been proposed that differentiate between different components of employability. DeFillippi and Arthur (1996) have argued that employability needs to be built on knowing what, that is career identity, knowing whom, that is social capital, and knowing how, that is personal adaptability and human capital. One model that is particularly relevant with respect to the issue of handling uncertainties is the model by Fugate and colleagues (2004). This model conceptualized employability within a broad framework of proactive behaviour in the face of uncertainty. Similarly to Arthur's definition, core components of employability are career identity, personal adaptability, and social and human capital. Career identity is understood as one's self-definition in the career context, which is inherently longitudinal because it involves making sense of the past and present as well as giving direction to one's future. Career identity therefore provides guidance for the personal adaptations needed to become and remain employable. Personal adaptability as manifested by personal characteristics such as propensity to learn, openness, optimism, internal locus of control, and generalized self-efficacy, helps people to identify and realize opportunities at work. Social capital defined by the size and strength of the social networks a person has access to, helps through increasing the amount of relevant information and the influence available to support job search and job applications. Human capital finally, concerns the competencies an individual has to offer and is determined by age, education, general abilities, work experience and training.

To date, there are more theoretical conceptualizations of employability than actual empirical tests of the proposed models. Two examples of empirical work have been provided by Wittekind (2007) who studied employability in several Swiss samples. She took investments in human capital by the employee and the employer, current level of job-related skills, willingness to be mobile with respect to job moves and the acquisition of new skills, knowledge of the labour market regarding job opportunities, and assessment of one`s own skills as antecedents of employability. Employability itself was defined as perceived chance of a job on internal or external labour markets (Forrier and Sels 2003). The main finding was that of all the antecedents included, only age, education, and current skill level were crucial determinants of employability. Similar strong evidence for the overriding relevance of human capital variables has been reported in other studies (e.g Berntson *et al.* 2006; Judge *et al.* 1999).

These findings are rather disconcerting as they indicate that people have little control over uncertain employment situations. Instead, personal characteristics that are difficult to change (education, skill level) or are not changeable at all (age) seem to be crucial for perceived employment options.

It has to be kept in mind that all these studies looked at perceptions of employability and not actual job opportunities or successful job transitions. One may argue that perceived employability is more relevant for developing a sense of control over an uncertain employment situation than actual opportunities. Moreover, current employment policies in companies do corroborate these results. Older employees are often sent into early retirement rather than offered new assignments or training opportunities. Employment statistics in many European countries show that not even the majority of employees work until the official retirement age. When competence requirements change in companies due to technological developments or new product and service portfolios, new employees with the required skills are hired rather than existing employees being trained to fill the new roles.

Given the emphasis that is frequently placed on social capital in the job search literature (*e.g.*, Granovetter 1995), the findings are also amazing. Knowing the right people appears to be of little importance compared to having the right skills and being young. This result may also be linked to the resistance in many companies against hiring people into a new job who come from different professional backgrounds (Higgins 2001; O'Mahony and Bechky 2006). Instead, companies prefer to hire younger people educated in the required field and train them further on the job. As Arthur and colleagues (2005) state, companies may not be as ready to support the self-directed career actor as they like to proclaim.

In Western societies, the emphasis on age and education in employment practices has been debated in recent years due to demographic changes. Faced with an aging work force, companies have to integrate older employees better and provide them with learning opportunities that, currently, still seem to be reserved for younger employees. The required changes in deeply rooted assumptions about the value, learning capabilities and adaptability of older individuals have been discussed within the framework of diversity management (*e.g.*, Ely and Thomas 2001).

8.3 Consequences for Human Resource Management

From the previous discussion, it is clear that human resource management needs to have at its disposal instruments for managing uncertainties in the employment relationship. Such instruments will be described under three headings: Strategic human resource management, as it relates to managing uncertainties regarding existing and required human resources; managing the psychological contract in order to support employees in handling uncertainties in the employment relationship; matching career orientations to company uncertainty profiles as a way of preparing employees for handling the uncertainties encountered.

8.3.1 Strategic Human Resource Management

The measures companies should take to handle uncertainties regarding existing and required human resources better are all related to the general notion of strategic human resource management. Strategic human resource management has been broadly defined as linking business needs with human resource practices and aligning the underlying human resource philosophy and policy with practices and processes (*e.g.*, Schuler 1992; Schuler and Jackson 2007). Uncertainties concern the prediction of human resource needs in relation to required skills, abilities, and knowledge, which are directly linked to strategic decisions on which products and services to provide to which markets via which processes. As external conditions change, these strategic decisions might have to be revised and with them, decisions on human resource needs. Uncertainties also concern the success in attracting and maintaining the targeted skills, abilities, and knowledge. Again, this is influenced by external factors such as educational systems, work force migration and mobility which affect the availability of individuals with the required characteristics on the labour market. Internal factors are also very relevant, concerning the attractiveness of the organization for prospective and current employees and the flexibility in personnel deployment and development in adjusting for changes in required skills and knowledge. The ensuing uncertainties and potentially inadequate means of control relate not only to human resource management practices, but also to strategic decision-making in general.

One interesting theoretical framework for supporting strategic human resource management in the face of these many uncertainties has been proposed and empirically tested by Lepak and Snell (1999, 2002). Starting from the basic question of make-or-buy regarding human resources, they suggest that organizations should decide on employment mode, employment relationship and human resources practices in view of the uniqueness and value of human capital for the company. If both uniqueness and value are high, that is firm-specific skills are needed that at the same time are vital to the competitive advantage of the organization, a traditional organization-focused employment relationship based on internal development of skills and high commitment should be aimed for. If both uniqueness and value of skills needed are low, contracting work out which fosters a transactional relationship based on compliance is recommended. These two examples already show that both strategies may be used within the same firm for different parts of the work force. For instance, development engineers might be employed in the first mode while basic maintenance work is contracted out following the second mode. If maintenance is safety-critical, however, the first mode should be used for those employees also (Hale *et al.* 1998). High value and low uniqueness put the firm in the difficult situation of having to try to retain someone whose skills can be put to use in many other companies as well. Therefore, the costs and benefits of the relationship have to be balanced and monitored with special care. Finally, high uniqueness and low value are best handled by forming alliances and collaborative partnership relationships as firm-specific skills are needed that, at the same time, are not valued enough to make internal development and retention worthwhile.

These different patterns of human resource management are suggested in view of handling the uncertainties related to attracting, retaining and developing unique and/or particularly valuable skills. The tightest control over resources is aimed for in the case of both high uniqueness and value, with the necessity of offering the employees concerned the highest possible benefit from the employment relationship. One may also look upon these strategies from the meta-perspective of how uncertainties are generally handled in an organization. A company strongly committed to a coping with uncertainties approach may use more commitment-based practices even for employees with less unique or valued skills.

8.3.2 Managing the Psychological Contract

On the operational level one important task of human resource management is to help employees handle their share of uncertainty in the employment relationship. This can be achieved by managing the psychological contract more systematically and by supporting the development of skills and competencies that foster internal and external employability. The latter is very directly linked to the strategic level, though, as it requires predictions of future skill needs.

Managing the psychological contract entails at least three elements: building realistic expectations in the selection and hiring process, monitoring and adapting expectations if needed, and synchronizing the different channels used to express company expectations and inducement (*i.e.,* formal company communication, human resources communication, and supervisor-employee interaction). The importance of consistent organizational conditions in terms of expressed values and practices has been demonstrated by Lee and colleagues (2004) in their study on building the psychological safety needed for handling uncertainties related to the introduction of new technology in a well-established work process. An interesting additional point regarding company communication in particular was introduced by Clampitt and Williams (2005), who argued that the (mis-)match between employee preferences for handling uncertainty and the perceived approach to uncertainty management in the organization will affect the effectiveness of communication and sensemaking.

Increasingly, organizations use employer value propositions to describe the inducements they offer to their employees and to synchronize internal and external communication regarding these inducements during the different phases of employment. In Example 8.1., the development and use of an employer value proposition at Novartis, along with the human resource management practices related to the continuous assessment and readjustment of expectations, company inducements and employee contributions are described. This is a good illustration of the processes needed for explicitly managing the psychological contract in the different stages of the employment relationship.

Example 8.1. Managing the psychological contract at Novartis

(1) Determine contents and type of contract as part of employer branding

At Novartis, very deliberate efforts have been made to define employer expectations and inducements as part of employer branding. Employer inducements have been formulated in terms of an Employer Value Proposition (EVP), which is derived from the employer brand "world`s most respected health care company" and aims to define a "brand contract". The EVP "distils the essence of what an employer offers its employees". In particular it aims to attract employees by showing what is special about working at Novartis, and retaining employees by living up to the promises made, thereby also measuring the performance as an employer. Core elements of the EVP at Novartis are: innovation for improving lives; global leader in healthcare; working with the best in the field; diversity; inspirational, high performance, achievement-oriented; professional and personal progression; results-driven rewards. A short version of the EVP is used as the introduction to all recruitment marketing materials: "A global healthcare leader, Novartis has one of the most exciting product pipelines in the industry today. A pipeline of innovative medicines brought to life by diverse, talented, performance driven people. All of which makes us one of the most rewarding employers in our field."

Employer expectations are also partially built into the EVP with its strong emphasis on performance and professional progression. Additionally, Corporate Social Responsibility (CSR) principles and the Novartis Code of Conduct require law-abiding and ethical behaviour, loyalty, integrity, respect, fairness and good business practice. More specific expectations are formulated as part of the values and behaviours used to assess performance and will be described under point 2.

The EVP also implies a particular type of psychological contract as it covers all employees with some flexibility for variations based on geographical, role or business area requirements, in the sense of "one voice, local accents". It is intended to help build an enduring relationship between Novartis and its employees.

(2) Communicate employer expectations and inducements in a coherent manner across all stages of the employment relationship

At Novartis, the following instruments and processes are used.

Recruitment. The EVP is consistently used in all recruiting materials.

Induction. Along with the legal employment contract, employees receive the brochure Novartis and You: "This brochure contains all the important information about the conditions of employment, benefits and offers you can expect from Novartis, as your company. This brochure is not part of your employment contract. Novartis reserves the right to change the benefits described here or to adapt them to current business and market requirements." The brochure spells out offers as well as expectations in a very detailed manner. For example, under the heading of performance management and personal development, an open dialogue on performance issues is offered as well as expected, and company investment in training and development is promised.

Performance appraisal. The aim is to create a "shared understanding of behaviours linked to job success at Novartis". The basis for appraisals is the individual performance objectives agreed and the values and behaviours defined for Novartis as a whole. Examples are action orientation, customer focus, personal accountability, self-discipline, and mutual respect. These values and behaviours are known to everybody along with the more detailed behaviours used to assess them.

Development/promotion. Development is seen as an important element of the EVP. It is based on a systematic process of assessing talent, providing adequate job experience along specified career maps, and continuous learning through 360° feedback, mentoring and tailor-made courses.

(3) Provide opportunities for negotiating/readjusting reciprocal expectations, contributions, and inducements

Performance appraisals and employee surveys are the main instruments for evaluating the fit between company promises in the EVP and perceived company inducements, as well as between company expectations and employee contributions. Adjustments might not only concern individual employees, but also larger groups of employees. For instance, it was found that inadequate career development opportunities were a major reason for quitting employment at Novartis, especially in China and Russia. As a consequence, learning centres were established in Russia and China with a focus on leadership training.

8.3.3 Matching Career Orientations to Company Uncertainty Landscapes

In previous sections, empirical evidence was presented that seems to indicate the prevalence of traditional, loyalty and security oriented employment relationships despite the rhetoric of new psychological contracts and new careers. It appears that employees who embrace the new employment relationship may get frustrated because jobs and career paths in companies are more routine than company representatives want to admit. In many companies, human resource managers have begun to face the problem of not being able to offer new career opportunities to employees who have followed the call for increasing their employability by extensive external training. This problem hints at inadequate analyses of changes in human resource demands and, as a consequence, preparing employees for handling uncertainties that do not materialize. Instead of having employees shrink from uncertainties – which is the issue mainly discussed in human resource management – employees may feel that the psychological contract has been breached because the promise of uncertainties in terms of new challenges and learning opportunities has not been met.

As part of strategic human resource management, company uncertainty landscapes containing the types of uncertainties present in different organizational units and jobs within these units should be established, using the framework presented in Chapter 2. This information can be used to match personal characteristics with the demands and opportunities present in particular jobs. These personal characteristics may then be translated into personality dispositions used in selection, such as the personal strategies used to handle uncertainties. For jobs that contain many uncertainties this is actually done when, for instance, future employees in research and development are questioned about their tolerance of failure. At the other extreme, there are jobs, such as monitoring automatic systems in process plants, where the issue is more one of staying alert despite high amounts of routine and to find people with a high tolerance to repetitious work.

Extending uncertainty landscapes to include projected changes and resulting development paths provides the possibility of identifying needs for employees with particular career orientations. For instance, when few changes are assumed to be required in the mid-term, people will be hired with more traditional career orientations. However, projections may turn out to be wrong so that even in presumably stable situations there should be some element of keeping people flexible and ready to learn through jobs into which learning opportunities have been deliberately integrated. When jobs are very routine and neither offer nor require major individual development, disengaged career orientations may be favoured. Finally, in highly volatile organizations, an independent career orientation probably provides the best match. Nonetheless, in these situations employees should also not be left completely to their own devices when planning their careers. Instead, support for career development should be offered which transcends organizational boundaries. A simple tool to evaluate career orientations is presented in Example 8.2. along with comparative data from Switzerland and Great Britain on the occurrence of different career orientations (see Table 8.2).

Example 8.2. Career orientations in Switzerland and Great Britain

As part of representative surveys of British and Swiss employees, career orientations were measured by nine questions (Gerber *et al.* 2008; see Table 8.2). The numbers behind the items in Table 8.2 indicate their relevance for particular types of career orientations. The more items with (1) are chosen by particular individuals, the more likely they have an *independent orientation*, the more items with (2) are chosen, the more likely is a *traditional-loyalty focused orientation*, the more items with (3) are chosen, the more likely is a *traditional-promotion focused orientation*, and the more items with (4) are chosen, the more likely is a *disengaged orientation.* A more exact assignment to the career orientations requires statistical analysis of the response patterns using latent class analysis (Gerber *et al.*, in press).

In the given example, striking differences between Switzerland and Great Britain, but also some differences between the German and French speaking parts of Switzerland were found. These difference concern, for instance, the importance of job security, which is much higher in Great Britain, and the centrality of work and organizational commitment, which are both particularly low in Great Britain and particularly high in the German-speaking part of Switzerland. Such differences are highly relevant for the human resource management in companies operating in the two countries or two parts of the country in the case of Switzerland. The differences also raise issues of diversity management, that is to what degree individual differences should be promoted and company instruments tailored to those differences.

8.4 Chapter Summary

In this chapter, uncertainties were discussed from the perspective of the individual employee. First, job design literature was summarized with respect to the demands on handling uncertainties in everyday work processes. Job autonomy, meaning involvement in decisions on goals and ways of achieving them, and job control, meaning the actual means to achieve goals in a particular work situation, were identified as crucial prerequisites for individual management of operational uncertainty. Secondly, uncertainties in the employment relationship through in-

Table 8.2. Career orientations in Switzerland and Great Britain

Looking ahead at your work life, which of the following would you choose ?	Great Britain (n=1000)	Switzerland German-speaking part (n=1007)	Switzerland French-speaking part (n=363)
Being employable in a range of jobs (1)*	27%	54%	42%
Having job security (2)	73%	46%	58%
Managing your own career (1)	80%	81%	81%
Having your organization manage your career for you (2)	20%	19%	19%
A short time in lots of organizations (1)	12%	24%	30%
A long time with one organization (2)	88%	76%	70%
A series of jobs at the same kind of level (4)	45%	50%	54%
Striving for promotion into more senior posts (3)	55%	50%	46%
Living for the present (4)	45%	46%	52%
Planning for the future (3)	55%	54%	48%
Work as marginal to your life (4)	52%	17%	37%
Work as central to your life (3)	48%	83%	63%
A career is not important to you (4)	42%	72%	64%
Career success is very important to you (3)	58%	28%	36%
Commitment to yourself and your career (1)	80%	48%	61%
Commitment to the organization (2)	20%	52%	39%
Spend what you've got and enjoy it (4)	47%	44%	47%
Save for the future (3)	53%	56%	53%

*Explanations for numbers given in Example 8.2.

creasing flexibility of working arrangements were discussed, demanding individual acceptance of less job security and more effort to stay employable. In particular, consequences of flexibility were described in terms of changing psychological contracts, new forms of more flexible and independent careers, and increasing demands on employability. Measures to be taken as part of human resource management were suggested. Better forecasts of skills needed in a particular company are required, which is a task of strategic human resource management. Employer and employee expectations, contributions and inducements within the framework of psychological contracts need to be monitored and adjusted continuously to help maintain a supportive employment relationship. Individual differences in career orientations should be measured more systematically and used to establish employment relationships geared towards individual resources and needs as related to the management of uncertainty.

9

Conclusion

I began this book by suggesting a framework for defining and analyzing uncertainties in organizations. Starting from the assumption that individual decision-makers and organizations aim to gain and maintain self-control, contents, sources, causes, and consequences of uncertainties were distinguished that may impede control by reducing the predictability and transparency of situations. In accordance with the prerogatives of a particular decision to be taken – be it about which supplier to choose, how much to rely on rules for internal coordination, whether to invest in automation, or how to handle production disturbances – this framework can be used to sketch an uncertainty landscape within which choices on reducing, maintaining or increasing uncertainties can be made. Minimizing versus coping with uncertainty were presented as two generic approaches to handling uncertainties. From the discussion of their advantages and disadvantages in relation to different contingencies the argument was developed that most of the time, organizations need to mix the two approaches in order to achieve an adequate balance between stability and flexibility. Various concepts and examples from different industries were provided to illustrate how different ways of handling uncertainties increase or decrease an organization's control capacity and thereby also, firm performance. In the subsequent chapters, the generic principles derived for managing uncertainty were applied to different organizational domains. In particular, coordination through routines and rules, collaborative planning within supply chains, design of automated systems, safety management, and human resource management were discussed. In this final chapter the aim is to show how the generic framework for managing uncertainty can be used to improve the compatibility of design decisions across a variety of organizational and technological issues, thereby supporting a more integrated management approach. Also, implications for research and practice will be suggested.

9.1 Integrative Management of Uncertainty: An Example

By no means I want to suggest yet another all-inclusive management system that is doomed to fail because it tries to be too many things at the same time, especially to be general, simple, and exact, which is an impossibility. I want, rather, to give an example of how decisions on managing uncertainty in different organizational domains affect each other and of how these interactions might be handled better by using the suggested process for making choices on reducing, maintaining, and increasing uncertainties across domains.

The example I will discuss can be traced all through the book – a railway company, which like all others is faced with many new uncertainties through technological developments, privatization, stiff competition, especially regarding carriage of freight, and growing capacity demands. The safety department in this railway company was charged with evaluating the effects of all these developments on the capabilities and needs of different groups of employees for performing their jobs effectively and safely. As a first step, the technological and organizational changes which are underway or planned for the next ten years were collated, highlighting three particularly important clusters of changes: increasing automation of train control, centralization of traffic control, and higher traffic density.

In two workshops with representatives from safety, quality management, infrastructure, train operation, and maintenance, the KOMPASS method was used to assess the effects of these changes on the task profiles for train drivers, signallers, shunters, and maintenance and construction personnel. Each task profile was analyzed in detail in relation to assumed changes in complexity and uncertainty through automation and task interdependencies. For shunting, maintenance and construction, these analyses showed an increase in uncertainties related to managing task interdependencies within more interlinked and more tightly planned work processes. For train drivers, the main impact is automation, which, in the long run will reduce train driving to mere supervisory control functions. Finally, for traffic controllers and signallers, it appeared that the integration of these two functions in central traffic control centres might lead to a new, more complex job profile for traffic controllers, and a less complex profile covering routine operation for the former signallers. Also, the central role of traffic control and of the changes in that function through centralization and automation for all other functions became very obvious, raising the question of whether impacts on these other functions need to be considered more in planning and implementing changes in traffic control.

Overall, the most significant concern that emerged was the growing difficulty of managing task interdependencies due to the greater centralization of traffic control, fewer buffers in resource planning, and loss of shared understanding of work processes through divisionalization of the organization. In order to address this concern, several measures were taken: more integrated training across company divisions, the development of a guideline for job and system design tailored to the needs of the different company divisions, and the development of a guideline for rule management in cooperation with the railways regulator. Additionally and most importantly, an annual risk assessment was introduced that will permit the monitoring of changes in the uncertainty landscape for different job functions and in (mis-) matches between requirements and capacity for handling those uncertainties.

Some of the work that provided the basis for developing guidelines for rule management was presented in Chapter 4. An important element of the guidelines as they now stand is a decision tree that helps to clarify the amount of uncertainty to be handled in a given work process, the possibilities for reducing that uncertainty, and the requirements for training and support by fairly open rules in case the uncertainty has to be maintained.

In Chapters 2 and 3, the coordination required between job functions for shunting a train onto its departure track was analyzed in detail, which is a good illustration of the intricate dependencies between tasks and job functions in railway operation. As was seen in this example, uncertainties can only be partially reduced and there is considerable potential for unduly transferring uncertainties to other job functions. The decision process proposed in Chapter 3 allows a systematic analysis of the uncertainties and their distribution which can be as focused as the example in order to redefine the division and coordination of certain tasks between different job functions or can be carried out on a more abstract level as part of the risk assessments introduced in the railway company.

The integrated training and the participatory development of common guidelines for rule management and job and system design across company divisions are important measures in themselves and address some of the issues discussed in Chapters 4 and 6, but they are also highly relevant for maintaining a shared culture. Drawing on the discussion in Chapter 7, culture is seen as a crucial coordination mechanism for dealing with high levels of uncertainties in the highly interlinked work processes in train operation and maintenance.

Given that the general increase in uncertainties and task interdependencies is partly grounded in the attempts to increase capacity by tighter traffic planning and centralized traffic control, planning processes themselves are also evaluated. Relevant issues are tightening versus loosening interdependencies and feedforward versus feedback control, as discussed in Chapter 5. Several pilot projects were started in order to experiment with different mixtures of central and local optimization of planning and operation, for instance, by having train drivers optimize speed with less restrictions on certain routes and by improving local coordination of time buffers contained in the train schedules in order to accommodate delays due to construction and maintenance work.

Some discussion has also begun in the railway company on the effects of job insecurity due to privatization and the highly competitive environment for rail operations generally. Instruments such as those suggested in Chapter 8 could help to evaluate existing employment relationships and create a new sense of security built on a better capacity to handle personal and organizational uncertainties.

At no point in the analyses undertaken in the railway company, was an explicit exploration of belief systems and their effects on perceived costs and benefits of the different ways of handling uncertainty carried out – as proposed in the decision process presented in Chapter 3. Rather, in the workshops and in the subsequent development of the various guidelines, differences in preferences and beliefs regarding effective organizational design became apparent, but were not dealt with explicitly. Instead of confronting the different views, broad participation in guideline development was sought as a means to further collective sensemaking and the building of shared belief systems. This may actually be a better way of addressing belief systems than trying to discuss them directly, especially in organizations with a rationalistic culture, as in this case.

9.2 Research Implications

In the previous chapters, current knowledge and research gaps in a variety of domains relevant for system and organizational design have been discussed. Instead of reiterating these discussions, just a few questions for future research are pointed out, which will hopefully stir some readers' curiosity sufficiently to instigate further explorations into the management of uncertainty.

As for the general framework for analyzing and managing uncertainty, the most important research need is to apply the framework in different settings and to test its theoretical foundations and practical use. Similarly, the different instruments proposed for improving rule management, collaborative planning, system design, safety management, and the psychological contract will have to be validated further.

Moreover, a number of more specific research questions have been raised, such as:

- Does the balance between stability and flexibility have to be achieved within each organizational subsystem or across subsystems?
- When are strategies aimed at deliberately increasing uncertainty appropriate?
- What is the impact of distinguishing between autonomy and control on job and system design?
- How can rules help to etsablish routines for adequately allocating cognitive resources in non-routine situations?
- When does the flexible use of a routine lead to a permanent change of the routine and the rules attached to it?
- Can task interdependence and autonomy be increased simultaneously in order to further collaborative planning?
- Does trust in technical systems develop through controlling the system or as a subsitute for control?
- How does admitting to the limits of control affect risk management and system design in practice?
- Can uncertainty be established as a moderator in the relationship between safety and autonomy?
- Why are more independent forms of employment relationships and careers so rare?

9.3 Practice Implications

Many instruments and procedures covering different aspects of the management of uncertainty have been proposed in this book. Furthermore, some recommendations concerning specific domains in organizational and system design have been made. These are summarized here once more to provide orientation for anyone interested in applying any of the elements of uncertainty management in their organization.

The book is organized around an analytic framework for understanding uncertainty in organizations. This framework includes causes, sources, contents, and consequences of uncertainty, as well as the different actors involved in the management of uncertainty. The overall aim is to help decision-makers in organizations to sketch an uncertainty landscape that supports informed decisions on managing these uncertainties so as to achieve desired outcomes. The framework's elements can be applied to analyze conditions of managing uncertainty from a strategic and an operational perspective. Moreover, the framework allows the adoption of a rationalistic, objective stance on managing uncertainty, but also the consideration of the impact of individual and collective enactment and sensemaking, which is of particular importance precisely in situations with high levels of uncertainty.

Minimizing versus coping with uncertainties are contrasted as generic approaches to uncertainty management whose appropriateness depends on various contingencies, most of all the amount and kinds of uncertainty with which an organization is faced. Balancing stability and flexibility is considered the ultimate goal of adequate management of uncertainty. In order to reach this goal, a decision process should be followed that systematically considers the costs and benefits of reducing, maintaining and increasing uncertainty in a particular context and relates them to individual and collective belief systems influencing the perception of these costs and benefits.

Regarding rules and routines as coordination mechanisms, the main aim is to establish rules and routines that allow uncertainty to be coped with while at the same time providing sufficient standardization to ease coordination demands. Instead of only taking the amount of rules as an indication of more or less standardization, different types of rules need to be deliberately designed. For this purpose, a typology of rules should be used that distinguishes between goal rules, process rules, and action rules.

Operational planning within and across organizations should be regarded as a resource for situated action, allowing actors to discover and exploit opportunities for goal attainment and goal adjustment as they arise. Collaborative planning is suggested as central to establishing and maintaining effective relationships within and across organizations, which permit a fair distribution of uncertainty across the different partners involved. Specific characteristics of collaborative planning are communication of anticipated events, knowledge of reference field characteristics, goal agreements, negotiation of alternatives, recognition of planning adequacy, monitoring and error diagnosis, coordination of opportunistic planning, and common reflection/decision on plan cancellation if necessary.

A major concern in the design of automated systems is control and accountability. The KOMPASS method supports managers and engineers in designing systems that are controllable by the human operator especially in those conditions where all other safeguards fail. However, with increasingly complex systems, design should be founded on the premise that there are limits to the controllability of technology. This approach will force system designers, the organizations operating the systems and regulatory institutions to take on responsibility for the use of technical systems whose complexity can no longer be mastered entirely by human operators.

Safety has long been managed, mainly by attempting to minimize uncertainties following the basic understanding of safety as minimal acceptable risk. Newer approaches to safety management aim to combine different ways of handling uncertainties in order to improve the balance between stability and flexibility in line with the notion of loose coupling. Safety culture is seen as important for achieving loose coupling because culture can serve as a "soft" mechanism for coordination and integration in a formally decentralized organization. The instrument suggested for evaluating and designing safety management and safety culture emphasizes the importance of assessing the fit between uncertainties and their management and of changing organization design to improve the fit if needed.

Finally, uncertainties are to be considered from the viewpoint of the individual employee. Job autonomy, meaning involvement in decisions on goals and ways of achieving them, and job control, meaning the actual means to achieve goals in a particular work situation, are identified as crucial prerequisites for individual management of operational uncertainty. Uncertainties in the employment relationship due to increased flexibility of working arrangements need to be addressed by forecasts of skill requirements as part of strategic human resource management, by continuous monitoring and adjustment of psychological contracts, and by catering for individual resources and needs regarding uncertainty management.

Taking a systematic approach to managing uncertainties across different organizational domains will prepare organizations better for a future that is bound to bring even more uncertainties.

References

Adler PS (1995) Interdepartmental interdependence and coordination: the case of the design/manufacturing interface. Organ Sci 6:147–167

Adler PS, Borys B (1996) Two types of bureaucracy: enabling and coercive. Adm Sci Q 41:6189

Adler PS, Goldoftas B, Levine DI (1999) Flexibility versus efficiency? A case study of model changeovers in the Toyota production system. Organ Sci 10:43–68

Allaire Y, Firsirotu ME (1989) Coping with strategic uncertainty. Sloan Manag Rev 30:7–16

Amalberti R (1992) Safety in process-control – an operator-centered point-of-view. Reliab Eng Syst Saf 38:99–108

Amalberti R (1993) Safety in flight operations. In: Wilpert B, Qvale T (eds) Reliablity and safety hazardous work situations. Approaches to analysis and design. Lawrence Erlbaum Hove UK, pp 171–194

Amalberti R (2001) The paradoxes of almost totally safe transportation systems. Saf Sci 37:109–126

Anand G, Ward PT (2004) Fit, flexibility and performance in manufacturing: coping with dynamic environments. Prod Op Manag 13:369–385

Argote L (1982) Input uncertainty and organizational coordination in hospital emergency units. Adm Sci Q 27:420–434

Argote L, Fichman M, Turner ME (1989) To centralize or not to centralize – the effects of uncertainty and threat on group-structure and performance. Organizat Behav Hum Decis Process 43:58–74

Arnold J (1997) Managing careers into the twentieth century. Paul Chapman, London

Arthur MB, Inkson K, Pringle JK (1999) The new careers: individual action and economic change. Sage, London

Arthur MB, Khapova SN, Wilderom CPM (2005) Career success in a boundaryless career world. J Organizat Behav 26:177–202

Arthur MB, Rousseau DM (1996) The boundaryless career: a new employment principle for a new organizational era. Oxford University Press, New York

Ashby WR (1956) An introduction to cybernetics. Chapman and Hall, London

Astley WG, Van de Ven AH (1983) Central perspectives and debates in organization theory. Adm Sci Q 28:245–273

Bagchi PK, Chun B, Skjoett-Larsen T, Soerensen LB (2005) Supply chain integration: a European survey. Int J Logist Manag 16:275–294

Bailey RW (1989) Human performance of engineering, 2nd ed. Prentice-Hall, London

Bainbridge L (1983) Ironies of automation. Automatica 19:775–779

Bandura A (2001) Social cognitive theory: an agentic perspective. Annu Rev Psychol 52:1–26

Baram M (2007) Liability and its influence on designing for product and process safety. Saf Sci 45:11–30

Barley SR (1986) Technology as an occasion for structuring: evidence from observations of CT scanners and the social order of radiology departments. Adm Sci Q 31:78–108

Barley SR, Kunda G (1992) Design and devotion – surges of rational and normative ideologies of control in managerial discourse. Adm Sci Q 37:363–399

Barley SR, Kunda G (2004) Gurus, hired guns, and warm bodies: itinerant experts in a knowledge economy. Princeton University Press, Princeton

Barratt M (2004) Understanding the meaning of collaboration in the supply chain. Supply Chain Manag Int J 9:30–42

Bax EH, Steijn BJ, De Witte MC (1998) Risk management at the shopfloor: the perception of formal rules in high-risk work situations. J Conting Crisis Manag 6:177–188

Becker MC (2004) Organizational routines: a review of the literature. Ind Corp Change 13:643–677

Becker MC, Knudsen T (2005) The role of routines in reducing pervasive uncertainty. J Bus Res 58:746–757

Beckert J (1999) Agency, entrepreneurs, and institutional change. The role of strategic choice and institutionalized practices in organizations. Organ Stud 20:777–799

Beckman CM, Haunschild PR, Phillips DJ (2004) Friends or strangers? Firm-specific uncertainty, market uncertainty, and network partner selection. Organ Sci 15:259–275

Bellotti V, Edwards K (2001) Intelligibility and accountability: human considerations in context-aware systems. J Hum-Comput Interact 16:193–212

Benner MJ, Tushman ML (2003) Exploitation, exploration, and process management: the productivity dilemma revisited. Acad Manag Rev 28:238–256

Berntson E, Marklund S, Sverke M (2006) Predicting perceived employability: human capital or labour market opportunities? Econ Ind Democr 27:223–244

Bigley G, Roberts KH (2001) The incident command system: high-reliability organizing for complex and volatile task environments. Acad Manag J 44:1281–1299

Birdi K, Clegg C, Patterson M, Robinson A, Stride CB, Wall TD, et al. (2008) The impact of human resource and operational management practices on company productivity: a longitudinal study. Pers Psychol 61:467–501

Boudreau MC, Robey D (2005) Enacting integrated information technology: a human agency perspective. Organ Sci 16:3–18

Bourdieu P (2005) The social structures of the economy. Polity Press, Cambridge
Bourgeois LJ (1985) Strategic goals, perceived uncertainty, and economic performance in volatile environments. Acad Manag J 28:548–573
Bourgeois LJ, Eisenhardt KM (1988) Strategic decision processes in high-velocity environments – four cases in the microcomputer industry. Manag Sci 34:816–835
Bourrier M (1998) Constructing organizational reliability: the problem of embeddedness and duality. In: Misumi J, et al. (eds) Nuclear safety: a human factors perspective. Taylor & Francis, London, pp 25–48
Boy GA (1998) Cognitive function analysis. Ablex, London
Brand FS, Jax K (2007) Focusing the meaning(s) of resilience: resilience as descriptive concept and boundary object. Ecol Soc 12(1):23. [online] URL: http://www.ecologyandsociety.org/vol12/iss1/art23/
Brehmer B (1992) Dynamic decision making – human control of complex systems. Acta Psychologica 81:211–241
Brophy DR (1998) Understanding, measuring, and enhancing collective creative problem-solving efforts. Creativity Res J 11:199–229
Brown SL, Eisenhardt KM (1995) Product development – past research, present findings, and future directions. Acad Manag Rev 20:343–378
Brown SL, Eisenhardt KM (1997) The art of continuous change: linking complexity theory and time-paced evolution in relentlessly shifting organizations. Adm Sci Q 42:1–34
Burgelman RA (2002) Strategy as a vector and the inertia of coevolutionary lock-in. Adm Sci Q 47:325–357
Burke MJ, Chan-Serafin S, Salvador R, Sarpy SA, Smith A (2008) The role of national culture and organizational climate in safety training effectiveness. Eur J Work Organizat Psychol 17:133–152
Burns T, Stalker GM (1961) The management of innovation. Tavistock Publications, London
Cacciabue PC (2004) Human error risk management for engineering systems: a methodology for design, safety assessment, accident investigation and training. Reliab Eng Syst Saf 83:229–240
Cassar V (2001) Violating psychological contract amongst Maltese public service employees: occurrence and relationships. J Manag Psychol 16:194–208
Cherns A (1976) Principles of sociotechnical design. Hum Relat 29:783–792
Cherns A (1987) Principles of sociotechnical design revisted. Hum Relat 40:153–162
Clampitt PG, Williams ML (2005) Conceptualizing and measuring how employees and organizations manage uncertainty. Commun Res Rep 22:315–324
Clegg S (1981) Organization and control. Adm Sci Q 26:545–562
Clegg SR, Courpasson D, Phillips N (2006) Power and organizations. Foundations for organizational science. Sage, London
Cohen SG, Ledford GE (1994) The effectiveness of self-managing teams – a quasi-experiment. Hum Relat 47:13–43
Comstock DE, Scott WR (1977) Technology and structure of subunits – distinguishing individual and workgroup effects. Adm Sci Q 22:177–202

Conway N, Briner RB (2002) Full-time versus part-time employees: understanding the links between work status, the psychological contract, and attitudes. J Vocat Behav 61:279–301

Conway N, Briner RB (2005) Understanding psychological contracts at work. Oxford University Press, Oxford

Coyle-Shapiro JA, Conway N (2004) The employment relationship through the lens of social exchange. In: Coyle-Shapiro JA, et al. (eds) The employment relationship: examining psychological and contextual perspectives. Oxford University Press, Oxford, pp 5–28

Coyle-Shapiro JA, Kessler I (2000) Consequences of the psychological contract for the employment relationship: a large scale survey. J Manag Stud 37:903–930

Crawford S, MacCarthy BL, Vernon C, Wilson JR (1999) Investigating the work of industrial schedulers through field study. Cogn Tech Work 1:63–77

Cropanzano R, Prehar CA, Chen PY (2002) Using social exchange theory to distinguish procedural from interactional justice. Group Organ Manag 27:324–351

Crowston K (1997) A coordination theory approach to organizational process design. Organ Sci 8:157–175

Crozier M (1964) The bureaucratic phenomenon. University of Chicago Press, Chicago

Daft RL, Lengel RH (1984) Information richness: a new approach to managerial behavior and organizational design. In: Cummings LL, Staw BM (eds) Research in organizational behavior 6. JAI Press, Homewood IL, pp 191–233

Daft RL, Macintosh N (1981) A tentative exploration into the amount and equivocality of information processing in organizational work units. Adm Sci Q 26:207–224

Danese P (2007) Designing CPFR collaborations: insights from seven case studies. Int J Op Prod Manag 27:181–204

Das JP, Kar BC, Parrila RK (1996) Cognitive planning. The psychological basis of intelligent behavior. Sage, New Delhi

de Vos A, Buyens D, Schalk R (2003) Psychological contract development during organizational socialization: adaptation to reality and the role of reciprocity. J Organizat Behav 24:537–558

Dean JW, Yoon SJ, Susman GI (1992) Advanced manufacturing technology and organization structure – empowerment or subordination. Organ Sci 3:203–229

Deci EL, Ryan RM (1985) The general causality orientations scale – self-determination in personality. J Res Pers 19:109–134

DeFillippi RJ, Arthur MB (1996) Boundaryless contexts and careers: a competency-based perspective. In: Arthur MB, Rousseau DM (eds) The boundaryless career: a new employment principle for a new organizational era. Oxford University Press, Oxford, pp 116–131

Dekker S (2003) Failure to adapt or adaptations that fail: contrasting models on procedures and safety. Appl Ergon 34:233–238

Dekker S, Hollnagel E (2004) Human factors and folk models. Cogn Tech Work 6:79–86

Dekker S, Woods DD (2002) MABA-MABA or abacadabra? Progress on human-automation coordination. Cogn Tech Work 4:240–244

Denison DR, Hooijberg R, Quinn RE (1995) Paradox and performance – toward a theory of behavioral complexity in managerial leadership. Organ Sci 6:524–540

DeSanctis G, Poole MS (1994) Capturing the complexity in advanced technology use – adaptive structuration theory. Organ Sci 5:121–147

Duncan RB (1972) Characteristics of organizational environments and perceived environmental uncertainty. Adm Sci Q 17:313–327

Eby LT, Butts M, Lockwood A (2003) Predictors of success in the era of the boundaryless career. J Organizat Behav 24:689–708

Edmondson AC, Bohmer RM, Pisano GP (2001) Disrupted routines: team learning and new technology implementation in hospitals. Adm Sci Q 46:685–716

Eisenhardt KM (1985) Control – organizational and economic approaches. Manag Sci 31:134–149

Eisenhardt KM (1989) Making fast strategic decisions in high-velocity environments. Acad Manag J 32:543–576

Eisenhardt KM, Martin JA (2000) Dynamic capabilities: what are they? Strateg Manag J 21:1105–1121

Eisenhardt KM, Tabrizi BN (1995) Accelerating adaptive processes – product innovation in the global computer industry. Adm Sci Q 40:84–110

Ely RJ, Thomas DA (2001) Cultural diversity at work: the effects of diversity perspectives on work group processes and outcomes. Adm Sci Q 46:229–273

Emery FE (1959) Characteristics of socio-technical systems. Tavistock Documents No. 527, London

Entin EE, Serfaty D (1999) Adaptive team coordination. Hum Factors 4:312–325

Faraj S, Sambamurthy V (2006) Leadership of information systems development projects. IEEE Trans Eng Manag 53:238–249

Faraj S, Xiao Y (2006) Coordination in fast response organizations. Manag Sci 52:1155–1169

Farrington-Darby T, Pickup L, Wilson JR (2005) Safety culture in railway maintenance. Saf Sci 43:39–60

Federal Aviation Administration (2007) Joint planning and development office: actions needed to reduce risks with the next generation air transportation system. Report Number: AV-2007-031. FAA, Washington, DC

Feldman MS, Pentland BT (2003) Reconceptualizing organizational routines as a source of flexibility and change. Adm Sci Q 48:94–118

Feldman SP (2004) The culture of objectivity: quantification, uncertainty, and the evaluation of risk at NASA. Hum Relat 57:691–718

Ferraro F, Pfeffer J, Sutton RI (2005) Economics language and assumptions: how theories can become self-fulfilling. Acad Manag Rev 30:8–24

Flanagan JC (1954) The critical incident technique. Psychol Bull 51:327–358

Flin R, Mearns K, O'Connor P, Bryden R (2000) Measuring safety climate: identifying the common features. Saf Sci 34:177–192

Folger R, Cropanzano R (1998) Organizational justice and human resource management. Sage, Thousand Oaks, CA

Folta TB (1998) Governance and uncertainty: the trade-off between administrative control and commitment. Strateg Manag J 19:1007–1028
Forrier A, Sels L (2003) The concept employability: a complex mosaic. Int J Hum Resour Dev Manag 3:102–124
Fugate M, Kinicki AJ, Ashfort BE (2004) Employability: a psycho-social construct, its dimensions, and applications. J Vocat Behav 65:14–38
Funke J, Glodowski AS (1990) Planen und Problemlösen: Überlegungen zur neuropsychologischen Diagnostik von Basiskompetenzen beim Planen. Z für Neuropsychol 1:139–148
Galbraith J (1973) Designing complex organizations. Addison-Wesley, Reading, MA
Galbraith JR (1974) Organization design – an information processing view. Interfaces 4:28–36
Gauthereau V, Hollnagel E (2005) Planning, control and adaption: a case study. Eur Manag J 23:118–131
Gerber M, Grote G (2007) How do psychological contracts and career orientations influence work outcomes? Small Group Meeting on Psychological Contracts EAWOP, Zurich
Gerber M, Wittekind A, Bannwart M, Grote G, Staffelbach B (2007) Schweizer HR-Barometer: psychologischer Vertrag und Arbeitsplatz(un)sicherheit. Verlag Neue Zürcher Zeitung, Zurich
Gerber M, Wittekind A, Conway N, Grote G (2008) Generalizability of career orientations: a comparative study in Switzerland and Great Britain. Academy of Management Conference, August 2008 Anaheim, CA
Gerber M, Wittekind A, Grote G, Staffelbach B (in press) Exploring types of career orientation: a latent class analysis approach. J Vocat Behav
Gerwin D (1993) Manufacturing flexibility – a strategic perspective. Manag Sci 39:395–410
Gilbert C, Amalberti R, Laroche H, Paries J (2007) Errors and failures: towards a new safety paradigm. J Risk Res 10:959–975
Gilbert CG (2005) Unbundling the structure of inertia: resource versus routine rigidity. Acad Manag J 48:741–763
Gilson LL, Mathieu JE, Shalley CE, Ruddy TM (2005) Creativity and standardization: complementary or conflicting drivers of team effectiveness? Acad Manag J 48:521–531
Gittell JH (2000) Organizing work to support relational co-ordination. Int J Hum Resour Manag 11:517–539
Gouldner AW (1957) Cosmopolitans and Locals – toward an analysis of latent social roles 1. Adm Sci Q 2:281–306
Gouldner AW (1960) The norm of reciprocity – a preliminary statement. Am Sociol Rev 25:161–178
Granovetter M (1995) Getting a job. University of Chicago Press, Chicago
Green SG, Welsh MA (1988) Cybernetics and dependence – reframing the control concept. Acad Manag Rev 13:287–301
Greenhouse S (2008) The big squeeze: tough times for the American worker. Knopf, New York

Grote G (1997) Autonomie und Kontrolle. Zur Gestaltung automatisierter und risikoreicher Systeme. vdf Hochschulverlag, Zurich

Grote G (2001) Planen in Organisationen: Forschungserfordernisse und –ansätze. In: Silbereisen RK, Reitzle M (eds) Psychologie 2000. Bericht über den 42. Kongress der Deutschen Gesellschaft für Psychologie in Jena. Pabst, Lengerich, pp 576–586

Grote G (2004a) Organizational measures for achieving loose coupling in high-risk systems: the importance of systematic rules management. Paper presented at the IEEE-Conference on Systems, Man, and Cybernetics. The Hague, October 2004

Grote G (2004b) Uncertainty management at the core of system design. Annu Rev Control 28:267–274

Grote G (2007a) Understanding and assessing safety culture through the lens of organizational management of uncertainty. Saf Sci 45:637–652

Grote G (2007b) Arbeits- und Prozesssicherheit. In: Schuler H, Sonntag Kh (eds) Handbuch der Arbeits- und Organisationspsychologie. Hogrefe, Göttingen, pp 155–162

Grote G (2008a) Diagnosis of safety culture: a replication and extension towards assessing "safe" organizational change processes. Saf Sci 46:450–460

Grote G (2008b) Rules management as source for loose coupling in high-risk systems. In: Hollnagel E, Nemeth C, Dekker S (eds) Remaining sensitive to the possibility of failure. Ashgate, London, pp 91–100

Grote G, Baitsch C (1991) Reciprocal effects between organizational culture and the implementation of an office communication-system – a case study. Behav Inf Technol 10:207–218

Grote G, Gerber M, Günter H, Wittekind A (2007) Content matters: different reactions to psychological contract breach by cosmopolitans and locals, 23rd EGOS Colloquium, 7 July 2007, Vienna University of Economics and Business Administration, Austria

Grote G, Künzler C (1996a) Sicherheitskultur, Arbeitsorganisation und Technikeinsatz. vdf Hochschulverlag, Zurich

Grote G, Künzler C (1996b) Safety culture and its reflections in job and organizational design: total safety management. Int J Environ Pollut 6:618–631

Grote G, Künzler C (2000) Diagnosis of safety culture in safety management audits. Saf Sci 34:131–150

Grote G, Raeder S (2009) Careers and identity in flexible working: do flexible identities fare better? Hum Relat 62:219–244

Grote G, Ryser C, Wäfler T, Windischer A, Weik S (2000) KOMPASS: a method for complementary function allocation in automated work systems. Int J Hum-Comput Stud 52:267–287

Grote G, Turner N, Wall TD (2003) The uncertain relationship between autonomy and safety. In: Elisburg D (ed) (Chair) Occupational safety. Symposium at the 5th Work, Stress, and Health Conference, Toronto, Canada

Grote G, Weichbrodt JC, Günter H, Zala-Mezö E, Künzle B (2009) Coordination in high-risk organisations: the need for flexible routines. Cogn Tech Work 11:17–27

Grote G, Zala-Mezö E, Gommes P (2004) The effects of different forms of coordination on coping with workload. In: Dietrich R, Childress TM (eds) Group interaction in high risk environments. Ashgate, Aldershot, UK, pp 39–55

Guest D, Conway N (2004) Employee wellbeing at work and the psychological contract. Research report for the CIPD Charted Institute of Personnel and Development, London

Guldenmund FW (2000) The nature of safety culture: a review of theory and research. Saf Sci 34:215–257

Guldenmund FW (2007) The use of questionnaires in safety culture research – an evaluation. Saf Sci 45:723–743

Günter H (2007) Collaborative planning in heterarchic supply networks. Unpublished PhD thesis. ETH Zurich, Zurich

Gupta AK, Smith KG, Shalley CE (2006) The interplay between exploration and exploitation. Acad Manag J 49:693–706

Guttman R, Greenbaum CW (1998) Facet theory: its development and current status. Eur Psychol 3:13–36

Hacker W (2001) Psychological activity theory. In: Smelser NJ (ed) International encyclopedia of the social and behavioral sciences. Elsevier, Amsterdam, pp 58–62

Hackman JR, Oldham GR (1976) Motivation through design of work – test of a theory. Organizat Behav Hum Perform 16:250–279

Hale AR, Heijer T (2006) Is resilience really necessary? The case of railways. In: Hollnagel E (ed) Resilience engineering: concepts and precepts. Ashgate, Aldershot, UK, pp 125–148

Hale AR, Heijer T, Koornneef F (2003) Management of safety rules: the case of railways. Saf Sci Monit 7:1–11

Hale AR, Heming BHJ, Smit K, Rodenburg FGT, van Leeuwen ND (1998) Evaluating safety in the management of maintenance activities in the chemical process industry. Saf Sci 28:21–44

Hale AR, Swuste P (1998) Safety rules: procedural freedom or action constraint? Saf Sci 29:163–177

Hall DT (2002) Careers in and out of organizations. Sage, Thousand Oaks, CA

Handfield RB, Nichols EL (1999) Introduction to supply chain management. Prentice-Hall, Englewood Cliffs, NJ

Hauß Y, Timpe KP (2002) Automatisierung und Unterstützung im Mensch-Maschine-System. In: Timpe KP, et al. (eds) Mensch-Maschine-Systemtechnik. Konzepte, Modellierung, Gestaltung, Evaluation, 2nd ed. Symposion Düsseldorf, pp 41–62

Hayes-Roth B, Hayes-Roth F (1979) A cognitive model of planning. Cogn Sci 3:275–310

Hendry K, Kiel GC (2004) The role of the board in firm strategy: integrating agency and organisational control perspectives. Corp Gov-Int Rev 12:500–520

Heyman GD, Dweck CS (1992) Achievement goals and intrinsic motivation: their relation and their role in adaptive motivation. Motiv Emot 16:231–247

Hickson DJ, Hinings CR, Lee CA, Schneck RE, Pennings JM (1971) Strategic contingencies theory of intraorganizational power. Adm Sci Q 16:216–229
Higgins MC (2001) Changing careers: the effects of social context. J Organizat Behav 22:595–618
Hiltrop JM (1995) The changing psychological contract: the human resource challenge of the 1990s. Eur Manag J 13:286–294
Hindmarsh J, Pilnick A (2002) The tacit order of teamwork: collaboration and embodied conduct in anesthesia. Sociol Q 43:139–164
Hinds P, Mortensen M (2005) Understanding conflict in geographically distributed teams: an empirical investigation. Organ Sci 16:290–307
Hinings CR, Hickson DJ, Pennings JM, Schneck RE (1974) Structural conditions of intraorganizational power. Adm Sci Q 19:22–44
Hoc JM (1988) Cognitive psychology of planning. Academic Press, London
Hoc JM (2000) From human-machine interaction to human-machine cooperation. Ergon 43:833–843
Hoc JM (2001) Towards a cognitive approach to human-machine cooperation in dynamic situations. Int J Hum-Comput Stud 54:509–540
Hoekstra JM (2001) Designing for safety, the free flight air traffic management concept, NLR-TP-2001-313. National Aerospace Laboratory (NLR), Amsterdam
Hoffmann VH, Trautmann T, Schneider M (2008) A taxonomy for regulatory uncertainty-application to the European Emission Trading Scheme. Environ Sci Policy 11:712–722
Hofmann DA, Stetzer A (1996) A cross-level investigation of factors influencing unsafe behaviors and accidents. Pers Psychol 49:307–339
Hofstede G (2001) Culture's consequences, 2nd ed. Sage, Thousand Oaks, CA
Hollnagel E (1998) CREAM – Cognitive Reliability and Error Analysis Methodology. Elsevier, Oxford
Hollnagel E (2003) Handbook of cognitive task design. Lawrence Erlbaum, Mahwah, NJ
Hollnagel E (2007) Flight decks and free flight: where are the system boundaries? Appl Ergon 38:409–416
Hollnagel E, Bye A (2000) Principles for modelling function allocation. Int J Hum-Comput Stud 52:253–265
Hollnagel E, Nemeth CP, Dekker S (2008) Resilience engineering perspectives – vol. 1: remaining senstitive to the possibility of failure. Ashgate, Aldershot, UK
Hollnagel E, Woods DD (2005) Joint cognitive systems – foundations of cognitive systems engineering. Taylor & Francis, London
Hollnagel E, Woods DD, Leveson N (2006) Resilience engineering: concepts and precepts. Ashgate, Aldershot, UK
Hopkins A (2009) Thinking about process safety indicators. Saf Sci 47:460–465
Hornung S, Rousseau DM, Glaser J (2008) Creating flexible work arrangements through idiosyncratic deals. J Appl Psychol 93:655–664
Howard-Grenville JA (2005) The persistence of flexible organizational routines: the role of agency and organizational context. Organ Sci 16:618–636

Hoyos CG, Gockeln R, Palecek H (1981) Handlungs-orientierte Gefährdungsanalysen an Unfallschwerpunkten der Stahlindustrie. Z Arbeitswiss 35:146–149

Hunt JG, Stelluto GE, Hooijberg R (2004) Toward new-wave organization creativity: beyond romance and analogy in the relationship between orchestra-conductor leadership and musician creativity. Leadersh Q 15:145–162

Husseini SMM, O'Brien C, Hosseini ST (2006) A method to enhance volume flexibility in JIT production control. Int J Prod Econ 104:653–665

Ibarra H (2003) Working identity. Unconventional strategies for reinventing your career. Harvard Business School Press, Boston

Inagaki T (2000) Situation-adaptive autonomy for time-critical takeoff decisions. Int J Model Simul 20:175–180

Jackson S, Wilson JR, MacCarthy BL (2004) A new model of scheduling in manufacturing: tasks, roles, and monitoring. Hum Factors 46:533–550

Janis IL (1982) Groupthink: Psychological studies of policy decisions, 2nd ed. Houghton-Mifflin, Boston

Jarillo JC (1988) On strategic networks. Strateg Manag J 9:31–41

Jauch LR, Kraft KL (1986) Strategic management of uncertainty. Acad Manag Rev 11:777–790

Jordan N (1963) Allocation of functions between men and machines in automated systems. J Appl Psychol 47:161–165

Judge TA, Higgins CA, Thoreson CJ, Barrick MR (1999) The big five personality traits, general mental ability, and career success across the life span. Pers Psychol 52:621–652

Jungermann H, Schütz A, Theissen A, Doefke C (1991) Determinanten, Korrelate und Konsequenzen der Beurteilung von Risiken für die eigene Gesundheit. Z Arbeits- Organisationspsychol 35:59–67

Kaiser RB, Lindberg JT, Craig SB (2007) Assessing the flexibility of managers: a comparison of methods. Int J Sel Assess 16:40–55

Kaplan S, Beinhocker ED (2003) The real value of strategic planning. Sloan Manag Rev 44:71–76

Karasek R (1979) Job demands, job decision latitude, and mental strain: implications for job redesign. Adm Sci Q 24:285–307

Karasek R, Theorell T (1990) Healthy work, stress, productivity, and the reconstruction of working life. Basic Books, New York

Katz-Navon T, Naveh E, Stern Z (2005) Safety climate in health care organizations: a multidimensional approach. Acad Manag J 48:1075–1089

Kaufmann FX (1970) Sicherheit als soziologisches und sozialpolitisches Phänomen. Emke, Stuttgart

Keller RT, Slocum JW, Susman GI (1974) Uncertainty and type of management system in continuous process organizations. Acad Manag J 17:56–68

Ketokivi M (2006) Elaborating the contingency theory of organizations: the case of manufacturing flexibility strategies. Prod Op Manag 15:215–228

Kirwan B, Hale AR, Hopkins A (2002) Changing regulation: controlling hazards in society. Pergamon, Oxford

Klein JA (1991) A reexamination of autonomy in light of new manufacturing practices. Hum Relat 44:21–38

Klein KJ, Sorra JS (1996) The challenge of innovation implementation. Acad Manag Rev 21:1055–1080

Klemola UM, Norros L (1997) Analysis of the clinical behaviour of anaesthetists: recognition of uncertainty as a basis for practice. Med Educ 31:449–456

Kornwachs K (1999) Bedingungen verantwortlichen Handelns. In: Timpe K-P, Rötting M (eds) Verantwortung und Führung in Mensch-Maschine-Systemen. Pro Universitate, Sinzheim, pp 51–79

Kotter JP (1973) Psychological contract – managing joining-up process. Calif Manag Rev 15:91–99

Kraut R, Steinfeld C, Chan AP, Butler B, Hoag A (1999) Coordination and virtualization: the role of electronic networks and personal relationships. Organ Sci 10:722–740

Kreiser P, Marino L (2002) Analyzing the historical development of the environmental uncertainty construct. Manag Decis 40:895–905

Kruglanksi AW, Webster DM (1996) Motivated closing of the mind: seizing and freezing. Psychol Rev 103:263–283

Künzler C (2001) Kompetenzfördernde Sicherheitskultur – Ein Ansatz zur ganzheitlichen Gestaltung risikoreicher Arbeitssysteme. Unpublished PhD thesis. Universität Zurich, Zurich

Langfred CW (2007) The downside of self-management: a longitudinal study of the effects of conflict on trust, autonomy, and task interdependence in self-managing teams. Acad Manag J 50:885–900

LaPorte TR, Consolini PM (1991) Working in practice but not wrong in theory: theoretical challenges of high reliability organisations. J Adm Res Theory 1:19–47

Lawrence PR, Lorsch J (1967) Organization and environment. Harvard University Press, Cambridge, MA

Lawton R (1998) Not working to rule: understanding procedural violations at work. Saf Sci 28:77–95

Lazarus RS (1999) Stress and emotion: a new synthesis. Free Association Books, London

Lee F, Edmondson AC, Thomke S, Worline M (2004) The mixed effects of inconsistency on experimentation in organizations. Organ Sci 15:310–326

Lee J, Moray N (1992) Trust, control strategies and allocation of function in human machine systems. Ergon 35:1243–1270

Lee JD, Moray N (1994) Trust, self confidence and operators' adaptation to automation. Int J Hum-Comput Stud 40:153–184

Lee JD, See KA (2004) Trust in automation: designing for appropriate reliance. Hum Factors 46:50–80

Leifer R, Mills PK (1996) An information processing approach for deciding upon control strategies and reducing control loss in emerging organizations. J Manag 22:113–137

Leontjew A (1981) The problem of activity in psychology. In: Wertsch JV (ed) The concept of activity in Soviet psychology. Sharpe, Armonk, NY

Lepak DP, Snell SA (1999) The human resource architecture: toward a theory of human capital allocation and development. Acad Manag Rev 24:31–48

Lepak DP, Snell SA (2002) Examining the human resource architecture: the relationships among human capital, employment, and human resource configurations. J Manag 28:517–543

Leplat J (1987) Occupational accident research and systems approach. In: Rasmussen J, Duncan K, Leplat J (eds) New technology and human error. Wiley, Chichester, pp 181–191

Leplat J (1998) About implementation of safety rules. Saf Sci 29:189–204

Levitt RE, Thomsen J, Christiansen TR, Kunz JC, Jin Y, Nass C (1999) Simulating project work processes and organizations: toward a micro-contingency theory of organizational design. Manag Sci 45:1479–1495

Lin WT, Shao BBM (2000) The relationship between user participation and system success: a simultaneous contingency approach. Inf Manag 37:283–295

Lind EA, Van den Bos K (2002) When fairness works: toward a general theory of uncertainty management. Res Organizat Behav 24:181–223

Lipshitz R, Strauss O (1997) Coping with uncertainty: a naturalistic decision-making analysis. Organizat Behav Hum Decis Process 69:149–163

Loch CH, Terwiesch C (2005) Rush and be wrong or wait and be late? A model of information in collaborative processes. Prod Op Manag 14:331–343

Lorenzoni G, Lipparini A (1999) The leveraging of interfirm relationships as a distinctive organizational capability: a longitudinal study. Strateg Manag J 20:317–338

Luhmann N (1979) Trust and power. Wiley, Chichester

Luhmann N (1988) Familarity, confidence, trust: problems and alternatives. In: Gambetta D (ed) Trust making and breaking cooperative relations. Blackwell, New York, pp 94–107

MacCrimmon KR, Wehrung DA (1990) Characteristics of risk taking executives. Manag Sci 36:422–435

Main Commission Aircraft Accident Investigation Warsaw (1994) Report on the Accident to Airbus A320-211 Aircraft in Warsaw on 14 September 1993, Warsaw, Poland, March 1994. Web version prepared by Peter Ladkin, URL: www.rvs.uni-bielefeld.de/publications/Incidents/DOCS/ComAndRep/Wasaw/warsaw-report.html

Majumdar SK, Marcus AA (2001) Rules versus discretion: the productivity consequences of flexible regulation. Acad Manag J 44:170–179

Manser T, Howard SK, Gaba DM (2008) Adaptive coordination in cardiac anaesthesia: a study of situational changes in coordination patterns using a new observation system. Ergon 51:1153–1178

March JG (1991) Exploration and exploitation in organizational learning. Organ Sci 2:71–87

March JG, Schulz M, Chou H, Zhou X (2000) The dynamics of rules: change in written organizational codes. Stanford University Press, Stanford, CA

Marchand A, Simard M, Carpentier-Roy MC, Ouellet F (1998) From a unidimensional to a bidimensional concept and measurement of workers' safety behavior. Scand J Work Environ Health 24:293–299

Markus ML, Robey D (1988) Information technology and organizational-change – causal-structure in theory and research. Manag Sci 34:583–598

Marris P (1996) The politics of uncertainty: attachment in private and public life. Routledge, London
Mathieu JE, Schulze W (2006) The influence of team knowledge and formal plans on episodic team process-performance relationships. Acad Manag J 49:605–619
McGuirl JM, Sarter NB (2006) Supporting trust calibration and the effective use of decision aids by presenting dynamic system confidence information. Hum Factors 48:656–665
McKay K (1992) Production planning and scheduling: a model for manufacturing decisions requiring judgement. Unpublished PhD thesis. University of Waterloo, Ontario, Canada
McKay K, Buzacott JA, Safayeni FR (1989) The scheduler‘s knowledge of uncertainty: the missing link. In: Browne J (ed) Knowledge based production management systems. North-Holland, Amsterdam, pp 171–189
McKeen JD, Guimaraes T, Wetherbe JC (1994) The relationship between user participation and user satisfaction – an investigation of four contigency factors. MIS Q 18:427–451
Mearns K, Whitaker SM, Flin R (2001) Benchmarking safety climate in hazardous environments: a longitudinal, interorganizational approach. Risk Anal 21:771–786
Mearns K, Whitaker SM, Flin R (2003) Safety climate, safety management practice and safety performance in offshore environments. Saf Sci 41:641–680
Meier K (2005) Stakeholder-Dialog über Pervasive Computing – überall und unsichtbar. Stiftung Risiko-Dialog, St. Gallen
Mesquita LF, Anand J, Brush TH (2008) Comparing the resource-based and relational views: knowledge transfer and spillover in vertical alliances. Strateg Manag J 29:913–941
Meyer AD, Tsui AS, Hinings CR (1993) Configurational approaches to organizational analysis. Acad Manag J 36:1175–1195
Miller D, Shamsie J (1999) Strategic responses to three kinds of uncertainty: product line simplicity at the Hollywood film studios. J Manag 25:97–116
Miller GA, Galanter E, Pribram K (1960) Plans and the structure of behavior. Holt, New York
Miller KD (1992) A framework for integrated risk management in international-business. J Int Bus Stud 23:311–331
Milliken FJ (1987) Three types of perceived uncertainty about the environment – state, effect, and response uncertainty. Acad Manag Rev 12:133–143
Mirvis PH, Hall DT (1994) Psychological success and the boundaryless career. J Organizat Behav 15:365–380
Mitchell VL, Nault BR (2007) Cooperative planning, uncertainty, and managerial control in concurrent design. Manag Sci 53:375–389
Moray N, Inagaki T, Itoh M (2000) Adaptive automation, trust, and self-confidence in fault management of time-critical tasks. J Exp Psychol-Appl 6:44–58
Morgan G (1986) Images of organizations. Sage, Beverly Hills, CA
Morrison EW (2006) Doing the job well: an investigation of pro-social rule breaking. J Manag 32:5–28

Morrison EW, Robinson SL (1997) When employees feel betrayed: a model of how psychological contract violation develops. Acad Manag Rev 22:226–256
Muir BM (1994) Trust in automation: 1. Theoretical issues in the study of trust and human intervention in automated systems. Ergon 37:1905–1922
Müller S, Brauner C, Grote G, Künzler C (1998) Safety culture – a reflection of risk awareness. Swiss Re, Zurich
Mumford MD, Schultz RA, Osburn HK (2002) Planning in organizations: performance as a multi-level phenomenon. In: Yammario FJ, Dansereau F (eds) The many faces of multi-level issues (Research in multi-level issues, vol 1). Elsevier, Oxford, UK, pp 3–65
Nassimbeni G (1998) Network structures and co-ordination mechanisms – a taxonomy. Int J Op Prod Manag 18:538–554
Naveh E (2007) Formality and discretion in successful R&D projects. J Op Manag 25:110–125
Naveh E, Katz-Navon T, Stern Z (2005) Treatment errors in healthcare: a safety climate approach. Manag Sci 51:948–960
Neal A, Griffin MA, Hart PM (2000) The impact of organizational climate on safety climate and individual behavior. Saf Sci 34:99–109
Nelson RR, Winter SG (1982) An evolutionary theory of economic change. Harvard University Press, Cambridge, MA
Nicholson N (2005) Personality and domain-specific risk taking. J Risk Res 8:157–176
O'Mahony S, Bechky BA (2006) Stretchwork: managing the career progression paradox in external labor markets. Acad Manag J 49:918–941
O'Mahony S, Ferraro F (2007) The emergence of governance in an open source community. Acad Manag J 50:1079–1106
O'Reilly CA, Tushman ML (2004) The ambidextrous organisation. Harv Bus Rev 82:74–81
Oesterreich R, Resch M (1985) Zur Analyse arbeitsbezogener Kommunikation. Z Sozialisationsforsch Erziehungssoziol 5:271–290
Orlikowski WJ (1992) The duality of technology – rethinking the concept of technology in organizations. Organ Sci 3:398–427
Orlikowski WJ (2000) Using technology and constituting structures: a practice lens for studying technology in organizations. Organ Sci 11:404–428
Orton JD, Weick KE (1990) Loosely coupled systems – a reconceptualization. Acad Manag Rev 15:203–223
Ouchi WG (1979) A conceptual framework for the design of organizational control mechanisms. Manag Sci 25:833–848
Ouchi WG, Maguire MA (1975) Organizational control – two functions. Adm Sci Q 20:559–569
Palmer TB, Wiseman RM (1999) Decoupling risk taking from income stream uncertainty: a holistic model of risk. Strateg Manag J 20:1037–1062
Papadopoulos Y, McDermid JA (2001) Automated safety monitoring: a review and classification of methods. Int J Cond Monit Diagn Eng Manag 4:14–32
Parasuraman R, Riley V (1997) Humans and automation: use, misuse, disuse, abuse. Hum Factors 39:230–253

Parasuraman R, Sheridan TB, Wickens CD (2008) Situation awareness, mental workload, and trust in automation: viable, empirically supported cognitive engineering constructs. J Cogn Eng Decis Mak 2:140–160

Parker SK, Axtell CM (2001) Seeing another viewpoint: antecedents and outcomes of employee perspective taking. Acad Manag J 44:1085–1100

Parker SK, Axtell CM, Turner N (2001) Designing a safer workplace: importance of job autonomy, communication quality, and supportive supervisors. J Occup Health Psychol 6:211–228

Parker SK, Wall TD (1998) Job and work design: organizing work to promote well-being and effectiveness. Sage, Thousand Oaks, CA

Parsons T (1960) Structure and process in modern societies. Free Press, Glencoe, IL

Pasmore WA (1988) Designing effective organizations: the socio-technical systems perspective. Wiley, New York

Pasmore WA, Francis C, Haldeman J, Shani A (1982) Sociotechnical systems: a North American reflection on empirical studies of the seventies. Hum Relat 35:1179–1204

Perrow C (1967) Framework for comparative analysis of organizations. Am Sociol Rev 32:194–208

Perrow C (1983) The organizational context of human-factors engineering. Adm Sci Q 28:521–541

Perrow C (1984) Normal accidents: living with high-risk technologies. Basic Books, New York

Pfeffer J (1998) Seven practices of successful organizations. Calif Manag Rev 40:96–124

Pfeffer J, Salancik GR (1978) The external control of organization: a resource dependence perspective. Harper & Row, New York

Pfeffer J, Veiga JF (1999) Putting people first for organizational success. Acad Manag Exec 13:37–48

Pidgeon N (1998) Safety culture: key theoretical issues. Work Stress 12:202–216

Power M (2004) The risk management of everything: rethinking the politics of uncertainty. Demos, London

Quinn RE (1988) Beyond rational management: mastering the paradoxes and competing demands of high performance. Jossey-Bass, San Francisco

Quinn RE, Rohrbaugh J (1983) A spatial model of effectiveness criteria – towards a competing values approach to organizational analysis. Manag Sci 29:363–377

Raeder S, Grote G (2001) Flexibilität ersetzt Kontinuität. Veränderte psychologische Kontrakte und neue Formen persönlicher Identität. Arbeit – Z Arbeitsforsch, Arbeitsgestalt, Arbeitspolit 10:352–364

Raeder S, Grote G (2004) Fairness als Voraussetzung für die Tragfähigkeit psychologischer Verträge. In: Schreyögg G, et al (eds) Gerechtigkeit und Management (Managementforschung, Bd. 14) Gabler, Wiesbaden, pp 139–174

Rajaonah B, Anceaux F, Vienne F (2006) Study of driver trust during cooperation with adaptive cruise control. Trav Hum 69:99–127

Rasmussen J (1983) Skills, rules, and knowledge – signals, signs, and symbols, and other distinctions in human-performance models. IEEE Trans Syst Man Cybern 13:257–266

Rasmussen J (1997) Risk management in a dynamic society: a modelling problem. Saf Sci 27:183–213

Reason J (1997) Managing the risks of organizational accidents. Ashgate, Aldershot, UK

Reason J, Parker D, Lawton R (1998) Organizational controls and safety: the varieties of rule-related behaviour. J Occup Organizat Psychol 71:289–304

Reilly PA (1998) Balancing flexibility – meeting the interests of employer and employee. Eur J Work Organizat Psychol 7:7–22

Resch M (1988) Die Handlungsregulation geistiger Arbeit. Huber, Bern

Reynaud B (2005) The void at the heart of rules: routines in the context of rule-following. The case of the Paris Metro Workshop. Ind Corp Change 14:847–871

Ring PS, Van de Ven AH (1992) Structuring cooperative relationships between organizations. Strateg Manag J 13:483–498

Ring PS, Van de Ven AH (1994) Developmental processes of cooperative interorganizational relationships. Acad Manag Rev 19:90–118

Roberts KH (1993) New challenges to understanding organizations. MacMillan, New York

Robey D (1977) Computers and management structure – some empirical findings reexamined. Hum Relat 30:963–976

Robinson SL, Rousseau DM (1994) Violating the psychological contract: not the exception but the norm. J Organizat Behav 15:245–259

Rogoff B, Baker-Sennett J, Matusov E (1994) Considering the concept of planning. In: Haith MM, et al. (eds) The development of future-oriented processes. University of Chicago Press, Chicago, pp 353–373

Rogoff B, Gauvain M, Gardner W (1987) The development of children's skills in adjusting plans to circumstances. In: In: Friedman SL, et al. (eds) Blueprints for thinking: the role of planning in cognitive developement. Cambridge University Press, Cambridge, UK, pp 303–320

Rousseau DM (1989) Psychological and implied contracts in organizations. Empl Rights Responsib J 2:121–139

Rousseau DM (1990) New hire perceptions of their own and their employer's obligations: a study of psychological contracts. J Organizat Behav 11:389–400

Rousseau DM (1995) Psychological contracts in organizations: understanding written and unwritten agreements. Sage, Thousand Oaks, CA

Saari JT, Lathela J (1978) Job enrichment cause of increased accidents? Ind Eng 8:41–45

Salancik GR, Pfeffer J (1978) A social information-processing approach to job attitudes and task design. Adm Sci Q 23:224–253

Sanderson PM (1989) The human planning and scheduling role in advanced manufacturing systems – an emerging human factors domain. Hum Factors 31:635–666

Santoro MD, McGill JP (2005) The effect of uncertainty and asset co-specialization on governance in biotechnology alliances. Strateg Manag J 26:1261–1269

Sarasvathy SD (2001) Causation and effectuation: toward a theoretical shift from economic inevitability to entrepreneurial contingency. Acad Manag Rev 26:243–263

Satyanarayanan M (2001) Pervasive computing: vision and challenges. IEEE Pers Commun 8:10–17

Schein EH (1985) Organizational culture and leadership. Jossey-Bass, San Francisco

Scholnick EK, Friedman SL (1987) The planning construct in the psychological literature. In: Friedman SL, et al. (eds) Blueprints for thinking: the role of planning in cognitive development. Cambridge University Press, Cambridge, UK, pp 3–38

Schuler RS (1992) Strategic human resource management: linking people with the needs of the business. Organizat Dyn 21:18–32

Schuler RS, Jackson SE (2007) Strategic human resource management. Blackwell, Oxford, UK

Schwartz SH (1999) Cultural value value differences: some implications for work. Appl Psychol Int Rev 48:23–47

Scott WR (1981) Organizations: rational, natural and open systems. Prentice-Hall, Englewood Cliffs, NJ

SESAR consortium (2008). SESAR master plan. Eurocontrol, Brussels

Shannon HS, Mayr J, Haines T (1997) Overview of the relationship between organizational and workplace factors and injury rates. Saf Sci 26:201–217

Shaw MLG, Gaines BR (1996) Requirements acquisition. Softw Eng J 11:149–165

Shenhar AJ (2001) One size does not fit all projects: exploring classical contingency domains. Manag Sci 47:394–414

Sherali HD, Smith JC, Trani AA, Sale S (2000) National airspace sector occupancy and conflict analysis models for evaluating scenarios under the free-flight paradigm. Transp Sci 34:321–336

Sheridan TB (1987) Supervisory control. In: Salvendy G (ed) Handbook of human factors. Wiley, New York, pp 1243–1268

Shye S (1998) Modern facet theory: content design and measurement in behavioral research. Eur J Psychol Assess 14:160–171

Sitkin SB, Pablo AL (1992) Reconceptualizing the determinants of risk behavior. Acad Manag Rev 17:9–38

Sitkin SB, Sutcliffe KM, Schroeder RG (1994) Distinguishing control form learning in total quality management – a contingency perspective. Acad Manag Rev 19:537–564

Slocum JW, Sims HP (1980) A typology for integrating technology, organization, and job design. Hum Relat 33:193–212

Smith WK, Tushman ML (2005) Managing strategic contradictions: a top management model for managing innovation streams. Organ Sci 16:522–536

Sproull L, Kiesler S (1991) Computers, networks and work. Sci Am 265:116–123

Staehle WH (1991) Redundanz, Slack und lose Kopplung in Organisationen: eine Verschwendung von Ressourcen? In: Staehle WH, Sydow J (eds) Managementforschung 1. De Gruyter, Berlin, pp 313–345

Stevenson M, Spring M (2007) Flexibility from a supply chain perspective: definition and review. Int J Op Prod Manag 27:685–713

Stinchcombe AL (2001) When formality works: authority and abstraction in law and organizations. University of Chicago Press, Chicago

Straussberger S, Boy G, Barjou S, Figarol S, Salis F, Debernard S, et al. (2008) PAUSA for the future – a synthesis of phase 1. June 2008. Final Report

Suchman LA (1987) Plans and situated actions: the problem of human-machine communications. Cambridge University Press, Cambridge, UK

Suchman LA (2002) Located accountabilities in technology production. Scand J Inf Syst 14:91–105

Sutton RI, Kahn RL (1987) Prediciton, understanding, and control as antidotes to organizational stress. In: Lorsch JW (ed) Handbook of organizational behavior. Prentice-Hall, Englewood Cliffs, NJ, pp 272–283

Sydow J, Windeler A (1998) Organizing and evaluating interfirm networks: a structurationist perspective on network processes and effectiveness. Organ Sci 9:265–284

Taylor FW (1911) The principles of scientific management. Harper & Row, New York

Thompson JD (1967) Organizations in action. McGraw-Hill, New York

Timpe K-P, Jürgensohn T, Kolrep H (eds) (2002) Mensch-Maschine-Systemtechnik. Konzepte, Modellierung, Gestaltung, Evaluation, 2nd ed. Symposion, Düsseldorf

Trist E, Higgin C, Murray H, Pollock A (1963) Organizational choice. Tavistock Publications, London

Trist EL, Susman GI, Brown GR (1977) Experiment in autonomous working in an American underground coal-mine. Hum Relat 30:201–236

Tschan F (1995) Communication enhances small-group performance if it conforms to task requirements – the concept of ideal communication cycles. Basic Appl Soc Psychol 17:371–393

Tschan F, Semmer NK, Gautschi D, Hunziker P, Spychiger M, Marsch SU (2006) Leading to recovery: group performance and coordinative activities in medical emergency driven groups. Hum Perform 19:277–304

Turner N (2002) Self-managing teams and safety performance: an employee-level investigation. Unpublished PhD thesis. University of Sheffield, Sheffield

Tushman ML, Nadler DA (1978) Information processing as an integrating concept in organizational design. Acad Manag Rev 3:613–624

Tushman ML, O'Reilly CA (1996) Ambidextrous organizations: managing evolutionary and revolutionary change. Calif Manag Rev 38:8–29

Ulich E (1998) Arbeitspsychologie, 4th ed. vdf Hochschulverlag, Zurich

Van de Ven AH, Delbecq AL, Koenig R (1976) Determinants of coordination modes within organizations. Am Sociol Rev 41:322–338

van Muijen JJ, Koopman P, De Witte K, De Cock G, Susanj Z, Lemoine C, et al. (1999) Organizational culture: the focus questionnaire. Eur J Work Organizat Psychol 8:551–568

Vermesch P (1985) Donnés d'observation sur l'utilisation d'une consigne écrite: l'atomisation de l'action. Trav Hum 48:161–172

Victor B, Boynton A, Stephens-Jahng T (2000) The effective design of work under total quality management. Organ Sci 11:102–117

Vlaar PWL, Van den Bosch FAJ, Volberda HW (2007) On the evolution of trust, distrust, and formal coordination and control in interorganizational relationships – toward an integrative framework. Group Organ Manag 32:407–429

Volpert W (1994) Wider die Maschinenmodelle des Handelns. Aufsätze zur Handlungsregulationstheorie. Pabst, Lengerich

von Cranach M, Tschan F (2001) Psychology of action planning. In: Smelser NJ (ed) International encyclopedia of the social and behavioral sciences. Elsevier, Amsterdam, pp 41–45

von Krogh G (1998) Care in knowledge creation. Calif Manag Rev 40:133–153

Wäfler A (2002) Verteilt koordinierte Autonomie und Kontrolle-eine arbeitspsychologische Konzeptergänzung für die autonomieorientierte Analyse, Bewertung und Gestaltung der Produktionsplanung und -steuerung. Unpublished PhD thesis. ETH Zurich, Zurich

Wäfler T (2001) Distributed but coordinated. Third workshop on human performance in planning and scheduling, 11-13 July 2001, Saint John's, Newfoundland, Canada, pp 1–15

Wäfler T, Grote G, Windischer A, Ryser C (2003) KOMPASS: a method for complementary system design. In: Hollnagel E (ed) Handbook of cognitive task design. Lawrence Erlbaum, Mahwah, NJ, pp 477–502

Wagemann R, Gordon FM (2005) As the twig is bent: how group values shape emergent task interdependence in groups. Organ Sci 16:687–700

Wall TD, Corbett JM, Martin R, Clegg CW, Jackson PR (1990) Advanced manufacturing technology, work design, and performance – a change study. J Appl Psychol 75:691–697

Wall TD, Cordery JL, Clegg CW (2002) Empowerment, performance, and operational uncertainty: a theoretical integration. Appl Psychol-Int Rev 51:146–169

Wall TD, Jackson PR, Davids K (1992) Operator work design and robotics system performance – a serendipitous field-study. J Appl Psychol 77:353–362

Wall TD, Jackson PR, Mullarkey S, Parker SK (1996) The demands-control model of job strain: a more specific test. J Occup Organizat Psychol 69:153–166

Waller MJ (1999) The timing of adaptive group responses to nonroutine events. Acad Manag J 42:127–137

Waterson E, Older Gray MT, Clegg CW (2002) A sociotechnical method for designing work systems. Hum Factors 44:376–391

Weber M (1947) The theory of social and economic organisation. Oxford University Press, New York

Wehner T, Endres E, Clases C (1996) Hospitationen als Gestaltungs- und Lernfelder zwischenbetrieblicher Kooperationsbeziehungen. In: Endres E, Wehner T (eds) Zwischenbetriebliche Kooperation: die Gestaltung von Lieferbeziehungen. Psychologie Verlags Union, Weinheim, pp 87–104

Weick K (2001) Making sense of the organization. Blackwell, Oxford

Weick KE (1976) Educational organizations as loosely coupled systems. Adm Sci Q 21:1–19
Weick KE (1979) The social psychology of organizing, 2nd ed. Addison-Wesley, Reading, MA
Weick KE (1987) Organizational culture as a source of high-reliability. Calif Manag Rev 29:112–127
Weick KE (1995) Sensemaking in organizations. Sage, Thousand Oaks, CA
Weick KE, Roberts KH (1993) Collective mind in organizations – heedful interrelating on flight decks. Adm Sci Q 38:357–381
Weick KE, Sutcliffe KM (2001) Managing the unexpected. Jossey-Bass, San Francisco
Weick KE, Sutcliffe KM, Obstfeld D (1999) Organizing for high reliability: processes of collective mindfulness. Res Organizat Behav 21:81–123
Weitz E, Shenhav Y (2000) A longitudinal analysis of technical and organizational uncertainty in management theory. Organ Stud 21:243–266
Weyer J (2006) Modes of governance of hybrid systems: the mid-air collision at Überlingen and the impact of smart technology. Sci Technol Innov Stud 2:127–149
Wickens CD, Mavor A, Parasuraman R, McGee JM (1998) The future of air traffic control: human operators and automation. National Academy Press, Washington, DC
Wildavsky A (1988) Searching for safety. Transaction Books, New Brunswick, NJ
Windischer A (2003) Kooperatives Planen – Theoretische Herleitung und empirische Überprüfung von Merkmalen und Einflussgrössen kooperativer Planungstätigkeiten in der abteilungsübergreifenden Bedarfsplanung. Unpublished PhD thesis, University of Zurich, Zurich
Windischer A, Grote G (2003) Success factors for collaborative planning. In: Seuring S, et al. (eds) Strategy and organziation in supply chains Physica, Heidelberg, pp 131–146
Windischer A, Grote G, Mathier F, Meunier Martins S, Glardon R (2009) Characteristics and organizational constraints of collaborative planning. Cogn Tech Work 11:87–101
Wittekind A (2007) Employability: an empirical analysis of its antecedents and its relevance for employees in Switzerland. Unpublished PhD thesis, ETH Zurich, Zurich
Wittekind A, Bernard U, Gerber M, Grote G, Staffelbach B (2006) Schweizer HR-Barometer: psychologischer Vertrag und Karriereorientierung. Verlag Neue Zürcher Zeitung, Zurich
Wittenbaum GM, Stasser G, Merry CJ (1996) Tacit coordination in anticipation of small group task completion. J Exp Soc Psychol 32:129–152
Woods DD (1995) The alarm problem and directed attention in dynamic fault management. Ergon 38:2371–2393
Woods DD (2006) Essential charateristics of resilience for organizations. In: Hollnagel E, et al. (eds) Resilience engineering: concepts and precepts. Ashgate, Aldershot, UK, pp 69–76
Woods DD, Shattuck LG (2000) Distant supervision-local action given the potential for surprise. Cogn Tech Work 2:242–245

Wreathall J (2006) Properties of resilient organizations: an initial view. In: Hollnagel E, et al. (eds) Resilience engineering: concepts and precepts. Ashgate, Aldershot, UK, pp 275–285

Wright P, Pocock S, Fields B (1998) The prescription and practice of work on the flight deck In: Green TRG (ed) Proceedings of the 9th European Conference on Cognitive Ergonomics (ECCE9). EACE Press, Limerick, pp 37–42

Zala-Mezö E, Wacker J, Künzle B, Brüesch M, Grote G (2009) The influence of standardisation and task load on team coordination patterns during anaesthesia inductions. Qual Saf Health Care 18:127–130

Zohar D (1980) Safety climate in industrial-organizations – theoretical and applied implications. J Appl Psychol 65:96–102

Zohar D (2000) A group-level model of safety climate: testing the effect of group climate on microaccidents in manufacturing jobs. J Appl Psychol 85:587–596

Zölch M (2001) Koordination von Arbeitshandlungen in der Produktion. Huber, Bern

Zollo M, Winter SG (2002) Deliberate learning and the evolution of dynamic capabilities. Organ Sci 13:339–351

Zuckerman M, Kuhlman DM (2000) Personality and risk-taking: common biosocial factors. J Pers 68:999–1029

Index

Hamburg, Germany

MIX
Papier aus verantwortungsvollen Quellen
Paper from responsible sources
FSC® C105338

If you have any concerns about our products,
you can contact us on
ProductSafety@springernature.com

In case Publisher is established outside the EU,
the EU authorized representative is:
Springer Nature Customer Service Center GmbH
Europaplatz 3, 69115 Heidelberg, Germany

Printed by Libri Plureos GmbH
in Hamburg, Germany